Training Circular
No. 25-8

HEADQUARTERS
DEPARTMENT OF THE ARMY
Washington, DC, 5 April 2004

# TRAINING RANGES

Contents

---

DISTRIBUTION RESTRICTION: Approved for public release; distribution is unlimited.

*This publication supersedes TC 25-8, 15 March 2004.

## FIGURES

**ARMY RANGES**

**TABLES**

**Page**

# PREFACE

This circular provides information about Army ranges. With AR 210-21, *Army Ranges and Training Land Program,* this circular provides guidance for developing and operating Army ranges. It is a working guide for trainers, range and mobilization planners, engineers, coordinators, and mandated range project review boards at all levels of the Active Army, Army National Guard, and Army Reserve. It is the primary guide for installation and major Army command range development plans and for developing the Army Master Range Plan.

Army Regulation 385-63, *Policies and Procedures for Firing Ammunition for Training Target Practice and Combat,* provides criteria and instructions for developing surface danger zones for weapons fired on Army and Marine Corps ranges. The US Army Training Support Center is the executive agent for the Deputy Chief of Staff for Operations and Plans, Headquarters Department of the Army, for the Army Range and Training Land Program. As such, the Army Training Support Center is the proponent for this circular.

Submit comments and recommendations on Department of the Army Form 2028, (Recommended Changes to Publications and Blank Forms) directly to Commander, US Army Training Support Center, ATTN: ATIC-ATMD, Fort Eustis, Virginia 23604-5166.

**"The Division fought to train under realistic conditions in preparation for combat. The priorities were clear and subordinate leaders executed taining that produced a disciplined, lethal and flexible force capable of accomplishing any tactical task.**

**- The requirement for tough realistic training has not changed in the past 227 years. The Division lived under the "train as you fight" motto for the 12 months that preceded the war. The training proficiency, lethality, and maturity of the Division serves as an example for the Army to follow.**

**- Lessons Learned:**

- **Key leader and soldier stabilization produced a mature fighting force.**
- **Combined arms force on force and live-fire exercises prepared units at all levels for combat.**
- **Command and control training under offensive conditions at task force through division level was essential to 3ID(M)'s success."** ***Operation Iraqi Freedom, Third Infantry Division (Mechanized) "Rock of the Marne",*** **After Action Report Final Draft, 12 May 2003**

## Chapter 1
## Introduction

### 1-1. Overview

a. Effective live training, carried out to a high doctrinal standard, is the cornerstone of operational success. The training of the critical tasks that individual, crew, platoon and companies have to accomplish to be combat ready is directly related to the availability and capability of live-fire ranges and maneuver areas. The continued improvement of live-fire ranges and facilities is key to the development of the go-to-war skills. Live-fire ranges and facilities will be even more important for units in the future as they must be able to be deploy and be operational within ninety-six hours in a combat zone. The successful implementation of the business end of the Future Combat System (FCS) Unit of Action is when the sharing of information in the sensor array allows for precision targeting and grouping of precision fires for target kills. This capability can only be validated on a live fire range or complex.

b. The Army modernization of the combat force from the Current Force to the Future Force will increase the lethality and precision of weapon systems over greater distances. As these weapons are introduced and combined with a net centric digital command and control the achievement of overwhelming tactical dominance and control of the battlefield will be obtained. The precision of the weapon system will allow for the determinate of appropriate force, preventing unnecessary loss of life, as well as laying the foundation for the future operational environment.

### 1-2. Army Ranges

a. The Army's family of training ranges provides training opportunities to develop and improve soldier and team proficiency and competence in the use of sophisticated weaponry. Individual soldier proficiency and collective training ranges realistically portray combat conditions to mold the team into an effective fighting unit.

b. The computer-controlled ranges of today allow trainers to develop scenarios and control targets and battlefield simulation devices. This permits soldiers and units to practice mission essential tasks in a stressful environment. Computerized systems also provide performance feedback. After-action reviews (AAR), using data recorded during training, permits the commander to assess the unit's performance. The accurate feedback allows leaders to assess the mission status of their units and design training programs to overcome the identified shortcomings. The performance feedback highlights positive actions to reinforce correct procedures and to foster soldiers' confidence, enabling soldiers and leaders to recognize and correct their shortcomings.

## 1-3. Challenge

a. As weapon systems become more lethal and capable of delivering greater firepower over increased distances Army ranges must change. Current training ranges are required to support soldiers using their weapons through live-fire, subcaliber devices, and laser and simulation technology. The ranges of the future must serve as the focal point of training as we integrate the Live-Virtual-Constructive linkups and add digital command and control elements. At company level and below soldiers train and hone their combat skills in live-fire and maneuver. In an era of intense resource competition, each dollar spent to develop, mitigate, or restore training ranges must deliver the maximum return in effective training and combat readiness.

b. With greater firepower and maneuver capabilities, the requirement for range land will continue to grow. Acquiring additional range lands will be difficult, so sustaining our current range lands becomes a critical task. Challenges to range land acquisition include: encroachment of commercial and private development, protection of threatened and endangered species, loss of wetlands, prevention of soil, surface and ground water contamination, deterioration of air quality and noise effects.

## 1-4. Approaches

a. One approach to meet the training challenge is to develop ranges capable of supporting training for multiple purposes, weapons, and combined arms. For example, training on an automated field-fire (AFF) range and an automated record-fire (ARF) range can be provided on a single modified record-fire (MRF) range, a land savings when compared to construction of separate ranges. The qualification training range (QTR) provides a similar capability to combine many other types of ranges, see Figure D-5.

b. Risk management integration is integral to successful range design, construction, use and sustainment. Early application of risk management in the planning process provides the planners and decision makes with alternatives that reduce safety hazards, limit adverse environmental impact and minimize operational constraints. Information and data developed during the risk management process will facilitate and expedite the development of environmental impact documentation and accelerate the approval process.

## 1-5. Planning

a. Range development projects require careful, deliberate planning by a team who coordinates the Range and Training Land Program (RTLP) process in accordance with Army Regulation (AR) 210-21, *Army Ranges and Training Land Program,* and this circular. The members of this interdisciplinary team are collectively called "range planners." Range planners should seek innovative alternatives to fulfill training requirements. At the same time, they consider the need for ranges to support surge requirements on short notice in the event of mobilization. Command interest and support are essential to each step of the development process. Accurate, complete justification ensures that resource expenditures provide an optimum return.

b. The range planning team consists of:

(1) trainers

(2) installation real property master planners

(3) environmental and natural/cultural resource managers

(4) financial and manpower resource managers

(5) range officers

(6) safety and occupational health managers

(7) force developers

(8) facility engineers

## Chapter 2
## Range Modernization

### 2-1. The Range Modernization Program

a. The increased accuracy and lethality of weapons was one driving force to change Army ranges. The other was the need to train soldiers in more realistic environments. The recognition that urban combat is unavoidable has resulted in new types of ranges such as Live Fire Exercise Shoothouses, Urban Assault Course, Battle Area Complex (BAX), and Combined Arms Collective Training Facility (CACTF). As the proponent schools identify training requirements, the Army will continue to modify existing ranges and build new types of ranges as required. The focus for ranges will be interoperability, standardization, improved targetry, digital capability and multi-purpose utility.

b. The Department of the Army (DA), Office of the Deputy Chief of Staff, G3 (ODCS, G-3), manages the RTLP. The ODCS, G-3 coordinates training range activities with the U.S. Army Training and Doctrine Command (TRADOC), Corps of Engineers (COE), Army Materiel Command (AMC) IAW 350-38, *Training Device Policies and Management*, and other major Army commands (MACOM) to achieve operational readiness.

c. Army doctrine requires combined arms teamwork and synchronization. Units must train for wartime combined arms operations. Combined arms proficiency results from regular practice of combat missions and tasks in the live domain. It starts with the development of individual skills. Individual skills, when combined and practiced build unit proficiency from crew through brigade task force.

d. The modernization of Army ranges under the RTLP supports this doctrine. Range design and construction must support the development of soldier skills in individual weapons and crew-served weapon systems. The ranges also support unit training to standards established in Army Training and Evaluation Program (ARTEP) and Mission Training Plans (MTP). Multiple-use ranges meet these requirements and reduce construction and operating costs by permitting training with a variety of weapons on the same range. Many of these ranges support collective training for small units.

e. Computer technology can equip ranges with target remote control. These devices can depict realistic battlefield conditions under a variety of offensive and defensive scenarios. Computer-recorded hits and misses enable trainers to analyze performance, provide corrective instruction, and provide accurate AARs. Computer technology combines with other training devices to create stressful, challenging scenarios for soldiers to train as they will fight.

f. Prudent implementation of the standardization program:

(1) Establishes a family of ranges for compatible weapons to provide training of one or more well-defined requirements.

(2) Base range size and configuration on the units that the range will support. The quantity of firing positions and target emplacements can be tailored to user needs, making the most effective use of the range.

(3) Provides training to meet standard weapons qualification and sustainment training requirements.

(4) Permits commanders to assess combat readiness and prepare individuals and units for advanced targeted training.

(5) Fosters standard operating procedures leading to a common understanding of force employment.

(6) Provides accurate throughput capabilities of ranges for mobilization planning. This lets mobilization planners determine the number of ranges needed to meet training requirements.

g. Standard design manuals are available for ranges included in the RTLP. These manuals provide layouts of firing positions, lanes, and target arrangements as guidelines for range planners, architects, and construction contractors. The manuals provide flexibility for range development projects to adapt to local throughput, terrain, surface danger zone (SDZ) placement, and environment. These conditions may warrant variances in the number of lanes and targets and in the configuration of ranges from one location to another.

h. The RTLP established automated ranges both in the continental United States (CONUS) and Outside CONUS (OCONUS) and includes active Army, Army Reserve, and Army National Guard installations. The complexities of ranges are from the Automated Field-Fire (AFF) Range which provides entry-level familiarization with the M16-series rifle to the highly sophisticated Digital Multipurpose Range Complexes (DMPRC) that supports collective training at the platoon level. In some cases, DMPRCs support companies in combined arms live-fire exercises (CALFEX) when augmented with adjacent ranges, maneuver areas, and additional targetry.

**2-2. Future Outlook**

a. Weapon systems currently planned or under development will require land resources to include expanded SDZs. Costs of ordnance used for training and qualifying will likely increase. Increased operating tempos may incur greater costs. Increased reliance on weapon system training simulators, embedded training devices, and the tactical engagement simulation (TES) training system can avoid or offset many of these costs. Planning for range development and range upgrades must anticipate the impact of new weapon systems and training aids, devices, simulators, and simulations (TADSS).

b. Ranges will evolve to support both live-fire and TES using non-live-firing systems similar to the Multiple Integrated Laser Engagement System (MILES). Although the emphasis on multiple-use ranges will continue, constraints on land and fiscal resources will limit construction of large range complexes. Long-range, direct-fire gunnery training will concentrate in areas capable of containing projectiles in existing SDZs or with land available to expand SDZs. Where expansion is not possible, other answers must be found, such as shared use of ranges or use of non-firing training systems.

c. Target control and scoring will continue to improve through more powerful computer systems. These systems will provide more flexibility to program training scenarios and permit event-driven target acquisitions. Voice-recognition systems linked with range-control computers will permit trainers and instructors to play a direct role in activating targets to keep pace with the evolving battle situation. Range instrumentation systems, incorporating position-location technology, will identify units and track their relative location to targets. Improved hit sensors, thermal imagery, and round discriminators will provide more accurate scoring and feedback on shooter performance. Through-sight video systems will confirm gun-laying accuracy and provide additional performance feedback. Battlefield effects will improve as targets are equipped with "shootback" simulators. Casualty-assessment systems will determine the effects of indirect fire. Observer-controllers (OC) equipped with electronic clipboards containing training and evaluation outlines (T&EO) will provide more effective training performance feedback, and input to lessons learned.

# Chapter 3
# Training Areas and Centers

## 3-1. Introduction

a. Army training doctrine provides for the progressive achievement and sustainment of proficiency in individual and collective tasks. The Army classifies ranges and training areas by level of training from individual soldier qualification skills, through integrated live-fire and maneuver unit training, to large-scale force-on-force exercises using TADSS including progressive excellence in markmanship both in training areas and on ranges.

b. Three types of training areas support progressively higher levels of proficiency—local training areas (LTAs), major training areas (MTAs), and combat training centers (CTCs).

c. LTAs and MTAs support individual, crew and unit collective training using a combination of standard ranges that provide the flexibility to support a variety of weapons and weapon systems. Stationary and moving targets, remotely controlled and supplemented with battlefield weapons-effects simulators, provide opportunities to train under challenging conditions. The target arrays portray a threat environment. Units conduct training on LTAs facilities under their installation's direct purview. These are usually limited in size. Training areas and ranges at home station and Army schools use a combination of ranges and TADSS to develop soldier and crew proficiency and to exercise small-unit mission essential tasks.

d. CTCs offer stressful exercises for maneuver battalions and command and control headquarters. These exercises focus on missions and tasks that enable the unit to capitalize on its strengths and improve its performance.

e. Occasionally units in overseas locations arrange for maneuver rights areas (MRAs) with local jurisdictions and the MACOM. This provides opportunities for periodic maneuver training on land not directly controlled by the Army.

## 3-2. Local Training Areas (LTA)

a. For CONUS-based active units, homestation training for individual and crew weapons proficiency training typically occurs in the LTA. Local training area facilities allow familiarization, qualification, and sustainment training with minimum impact on resources for travel to and from other types of training areas. Training focuses on individual through platoon weapons proficiency and battalion maneuver requirements. Tactical engagement simulation systems and other weapon system training simulators support maneuver and live-fire exercises.

b. Training land and range availability at overseas LTAs and installations are usually more constrained than those available for CONUS-based active units. Reserve Component (RC) centers, armories, and weekend training sites, face limited LTAs as well. Training proceeds to the extent available resources will support including individual and limited collective weapons proficiency and small-unit maneuver training.

## 3-3. Major Training Areas (MTA)

a. An MTA usually has enough ranges and training land resources to support:

(1) Collective live-fire proficiency.

(2) Combined arms live-fire exercises.

b. The MTAs are usually geographically separate from LTAs. For example:

(1) CONUS – Yakima Firing Center, subinstallation and approximately 130 miles southeast of Fort Lewis, Washington.

(2) OCONUS – Grafenwoehr Training Area (GTA) and Hohenfels Training Area (HTA), Germany; Pohakaloa Training Area (PTA), Hawaii.

c. Units training at MTAs concentrate on large unit collective live-fire training (platoon through battalion) and maneuver training (battalion or brigade) according to doctrine and standards. This builds on the training proficiency achieved at LTAs. An MTA provides training to sustain lessons learned from CTC experiences that LTAs could not tactically or doctrinally accommodate. The size of some installations supports simultaneous individual weapons proficiency and both small and large unit collective training (e.g., Fort Hood, Texas).

### 3-4. Combat Training Centers (CTC)

a. The CTCs are the Army's premier training centers. Instrumentation systems collect data to support training feedback to units. The Army derives lessons learned from the collected data and uses the lessons learned to improve doctrine, training, leader development, organization, materiel, personnel, and facilities (DTLOMPF). The four CTCs are:

(1) National Training Center (NTC), Fort Irwin, California.

(2) Combat Maneuver Training Center (CMTC), GTA (live-fire) and HTA (maneuver), Germany.

(3) Joint Readiness Training Center (JRTC), Fort Polk, Louisiana.

(4) Battle Command Training Program (BCTP), Fort Leavenworth, Kansas.

b. The three maneuver CTCs (NTC, CMTC, and JRTC) focus on brigade task force training using a combination of live-fire ranges and maneuver training land. The Army's goal is for battalions to experience the rigorous and realistic training environment at a maneuver CTC during a commander's tour. Organic to these CTCs are:

(1) A dedicated opposing force unit.

(2) Observer-controllers.

(3) Instrumentation systems not available at LTA and MTA sites.

c. The maneuver CTCs require sufficient land to accommodate doctrinal fire and maneuver training using multiple scenarios over varied terrain. This allows rehabilitation of the land and precludes over-familiarization with the terrain, which detracts from training realism.

### 3-5. Training Area Functional Uses

The Army's force modernization program has increased the demand for collective training.

a. As a result, greater reliance must be placed on the use of LTAs for individual training and the initial stages of crew training. Local training areas include cantonment areas at installations, RC centers, and armories. Tactical engagements can be simulated using appropriate TADSS to overcome the space limitations required for live-fire exercises. Training at LTAs achieves and sustains soldier and crew proficiency and prepares for advanced collective training at MTAs and CTCs.

(1) The collocation of an MPRC, a multipurpose training range (MPTR), and an infantry squad battle course (ISBC) is desirable at an MTA. This accommodates CALFEX and other forms of combined arms training using a wide variety of weapon systems and units. Collocating these facilities affords the land area and target arrays required to train platoon through company live-fire engagements. Although primarily designed for M1 Abrams tank and M2/M3 Bradley/Cavalry Fighting Vehicle (BFV/CFV) training, all division weapon systems can be trained on the MPRC. The Stryker vehicle can train on a MPRC/digital MPRC (DMPRC) though the BAX is preferred. Air defense artillery, field artillery, and attack helicopters can support combined arms scenarios. The MPRC supports up to company-size light infantry, airborne, and air assault forces. The DMPRC enhances the capabilities of the standard MPRC. The DMPRC increases target count, maneuver area size, and supports the collection of digital data to support the assessment of training.

(2) The MTA maneuver areas provide limited battalion and brigade maneuver force-on-force capabilities.

b. Combat training centers provide an advanced collective training experience for combat, CS, and CSS units. In contrast to MTAs, CTCs have a dedicated opposing force (OPFOR) trained in threat tactics, observer controllers, and instrumentation systems. During force-on-force engagements, an instrumentation system collects, reports, and stores engagement data from TES-equipped participants. OC's use this information to prepare AARs and post exercise take home packages (THP). Arrays of automated, radio-controlled targets are distributed over a 5-kilometer by 10-kilometer area for the offensive and defensive live-fire ranges. On the offensive live-fire range, the battalion task force employs maneuver and live fire to attack a defending force represented by an appropriate threat target array. On the defensive range, the task force engages a target array that represents multiple threats.

(1) The NTC provides CONUS-based maneuver battalion task forces the experience of a demanding high-intensity force-on-force engagement against an OPFOR organized and trained to replicate a threat motorized rifle regiment. During NTC rotations, battalion task forces also train on offensive and defensive live-fire ranges using organic and supporting direct- and indirect-fire weapon systems, artillery, helicopter gunship, and U.S. Air Force (USAF) close-air support.

(2) The JRTC provides training for contingency forces (light infantry, ranger, airborne, air assault, and special operations) in joint operations. Scenarios include forced and non-forced entry and special operations missions for low- and mid-intensity conflicts worldwide. These scenarios employ elements of the USAF Air Combat Command and Air Mobility Command. The JRTC supports maneuver in day and night exercises. The well-trained OPFOR represents a threat based on the scenario. Planned exercises include urban operations, small boat operations, river crossing operations, and offensive air support. The instrumentation system supports a two-battalion brigade task force with its supporting CS and CSS units.

(3) Combat forces in U.S. Army Europe (USAREUR) conduct combined arms, force-on-force training at the CMTC. The CMTC training takes place at two locations. Live-fire training takes place at GTA. GTA provides MPRC, MPTR and ISBC training ranges. HTA supports maneuver training. Units training at HTA use the TES training system with a mature instrumentation system. Infantry/armor battalions and cavalry squadrons train at the CMTC annually. A well-trained OPFOR portrays major elements of a scenario driven enemy force and are trained in the appropriate threat tactics and doctrine. The OPFOR provides opportunities for realistic exercises from low intensity conflicts to full combat. The CMTC instrumentation systems support exercise control and collects real time position/location and engagement data. This data is used to conduct AARs and prepare THPs.

# Chapter 4
## Range Requirements

### 4-1. Introduction

a. Training ranges are major contributors to keeping the Army ready to accomplish its missions. They also represent a considerable investment in time, land, and fiscal resources. Once a range training need is established, range planners must examine other alternatives to a new range such as:

(1) The required range may be available at another installation.

(2) An existing range might be converted to support multiple training requirements.

(3) A simulation, subcaliber training device, or reduced-scale range might satisfy the need.

(4) Existing ranges might increase operational hours.

b. Clearly defined requirements for individual, collective, or combined arms training must support decisions to initiate range development projects.

c. This chapter provides guidance on assessing needs for range development. AR 210-21 establishes the requirement for this assessment and lists many of the factors that require consideration.

d. Throughput is an important factor in deciding whether to undertake a major expenditure for range modification or construction. Required throughput is the number of individuals, crews, or units requiring a particular type of training. Throughput capacity is the number of soldiers, crews, or units a given range can support during a specific period. Throughput calculations help evaluate alternatives and determine how many firing points, lanes, or ranges are needed to satisfy a training requirement. They must include both peacetime and mobilization requirements.

e. This chapter presents formulae for calculating throughput and examples of applying the formulae to different weapon systems. It also provides a matrix of standard ranges. Appendix A provides tables to help unit trainers select appropriate ranges for training and weapons qualification. It further identifies ranges for the current and future forces including the SBCT.

### 4-2. Determining Training Requirements

a. Field Manuals (FM) 7.0, *Training the Force* and FM 7.1, *Battle Focused Training* are primary references for determining training requirements. As described in the FMs, battle focus enables unit leaders to develop a Mission Essential Task List (METL). The METL comes from the unit's war plans and external directives. It links the unit's collective mission essential tasks with the individual tasks that support them. It is also a checklist to identify requirements for collective training of these tasks.

b. Applicable ARTEP MTPs identify conditions and standards for collective training tasks. The MTPs show the relationships among wartime missions, collective tasks, battle drills, and individual tasks. They give examples and guidance for developing situational training exercises (STX) based on mission requirements. The MTPs provide training criteria for critical wartime mission tasks. For additional sources of individual and collective training objectives, see:

(1) DA Pamphlet (Pam) 350-38, *Standards in Weapons Training.*

(2) Soldier's manuals.

(3) Applicable gunnery manuals.

(4) Mobilization plans.

(5) Force modernization plans.

(6) General defense plans.

(7) Army Regulations.

**4-3. Assessing Range Needs**

a. DA Pamphlet 350-38 provides weapon system qualification standards and suggested training strategies to attain and sustain the standards. It establishes prescribed standards and training resources. For each weapon system the pamphlet outlines the number of rounds allocated per training event and the frequency of events for each weapon.

b. To assess range needs, match the METL, individual and collective task training objectives, and associated training range requirements to the suitable weapon system. DA Pamphlet 350-38 provides frequency requirements for each event and can be accessed from the Army Training Support Center (ATSC) website at http://www.atsc.army.mil/atmd/STRAC/stracweb.asp

c. In selecting ranges, consider:

(1) The number of troops and weapon systems at the installation.

(2) Range throughput capacity.

(3) External training support requirements; for example, those of the RC.

(4) Availability of Army Target System (ATS) ranges to support the gunnery tables and tasks outlined in weapon system field manuals.

(5) Types of ATS targets available.

(6) Overlapping Surface Danger Zones (SDZs).

(7) Environmental restrictions.

(8) Availability of training devices, simulations, and subcaliber munitions.

(9) Installation priorities.

d. Factors to consider when assessing range needs include:

(1) Unit METL.

(2) Weapon system quantities.

(3) Required gunnery qualification frequencies.

(4) STRAC standards.

(5) Availability and throughput capacity of ranges.

e. This assessment may identify a requirement for a new range or an upgrade to an existing range.

## 4-4. Range Availability

Subtract scheduled and unscheduled maintenance days, holidays, and inclement weather days from 365 calendar days to determine range availability. Master training schedules for tenant and regional units will impact range availability and must be considered by range managers. Table 4-1 indicates typical range availability.

**Table 4-1. Typical range availability (days)**

| Factors | AFF/ARF | MRF | MPTR | MPRC | DMPRC |
|---|---|---|---|---|---|
| Scheduled Maintenance | 22 | 25 | 55 | 59 | 59 |
| Unscheduled Maintenance | 16 | 11 | 15 | 17 | 17 |
| Holidays | 10 | 10 | 10 | 10 | 10 |
| Inclement Weather | 5 | 5 | 5 | 5 | 5 |
| Total Non-Available Days | 53 | 51 | 85 | 91 | 91 |
| Available Days | 312 | 314 | 280 | 274 | 274 |

## 4-5. Scheduling And Utilization

a. The Army's goal is to schedule ranges to be used 80 percent of the days available for training. For example, the MPTR shown in Table 4-1 should be scheduled for use at least 224 days (80% of 280 days) annually. Use of existing facilities is critical to the requirements identification and needs assessment phase discussed in AR 210-21. The Army's utilization goal for ranges and training land is 90 percent of scheduled days. Therefore, if the MPTR in Table 4-1 is scheduled for 224 days, it should be used at least 202 days of the year.

b. When calculating range requirements for a unit, consider:

(1) The number of training elements (e.g., soldiers, crews, and squads) assigned to the unit.

(2) The number of lanes available on the range.

(3) The average time required for one training element to complete a course of firing.

c. For example, experienced soldiers in an infantry company may have a high level of marksmanship proficiency completing the course of firing with greater efficiency than soldiers in other types of units. The analysis described below estimates both day and night firing to require 30 minutes per soldier. Although night firing requires more time because of operations in the dark, other factors compensate. Fewer rounds are fired at night (30 versus 40 for day firing), and the distances fired are shorter (25 and 50 meters versus 50 to 300 meters for day firing).

d. Calculate the time required to cycle one 200-soldier company through an MRF range as follows:

NUMBER OF LANES: 16

NUMBER OF SOLDIERS IN COMPANY: 200

TIME REQUIRED FOR ONE SOLDIER TO COMPLETE FIRING: 0.5 HOUR

NUMBER OF SOLDIERS PER HOUR = 16 ÷ 0.5 = 32

NUMBER OF HOURS = 200 ÷ 32 = 6.25 HOURS PER COMPANY

e. Calculate the total time required for zeroing, sustainment firing, and qualification firing by one company during the course of a year as follows:

**Zeroing:**

NUMBER OF LANES: 110

NUMBER OF SOLDIERS IN COMPANY: 200

TIME REQUIRED FOR ONE SOLDIER TO COMPLETE FIRING: 0.5 HOUR

FREQUENCY OF ZEROING PER YEAR (DAY ONLY): 2

NUMBER OF SOLDIERS PER HOUR = 100 ÷ 0.5 = 200

NUMBER OF HOURS = 2(200 ÷ 200) = 2 HOURS PER COMPANY PER YEAR

**Sustainment:**

NUMBER OF LANES: 16

NUMBER OF SOLDIERS IN COMPANY: 200

TIME REQUIRED FOR ONE SOLDIER TO COMPLETE FIRING: 0.5 HOUR

FREQUENCY OF SUSTAINMENT FIRING PER YEAR: 2 DAY AND 2 NIGHT

NUMBER OF SOLDIERS PER HOUR = 16 ÷ 0.5 = 32

NUMBER OF HOURS = 2(200 ÷ 32) = 12.5 DAY HOURS AND 12.5 NIGHT HOURS PER COMPANY PER YEAR

**Qualification:**

SAME FACTORS AS FOR SUSTAINMENT

NUMBER OF HOURS = 12.5 DAY AND 12.5 NIGHT HOURS PER YEAR

**Total time (hours) required for a single company to complete M16-series/M4 firing per year:**

| | | |
|---|---|---|
| ZEROING | 2.0 DAY HOURS | |
| SUSTAINMENT | 12.5 DAY HOURS | 12.5 NIGHT HOURS |
| QUALIFICATION | 12.5 DAY HOURS | 12.5 NIGHT HOURS |
| Total | 27.0 DAY HOURS | 25.0 NIGHT HOURS |

f. Given the foregoing, the annual range requirement (days) for M16-series/M4 rifle training for one company is 3.4 (27.0 hours ÷ 8 hours per day) days and 3.1 (25.0 hours ÷ 8 hours per night) nights.

g. The process shown above can be used to calculate usage requirements for other weapon systems, training requirements, and training elements.

**4-6. Quantifying Range Requirements**

a. Department of the Army Pamphlet 350-38 identifies weapon system training requirements, frequency of range events, and ammunition resources allocated to training. Using the formula in the throughput calculations paragraph below, installation range personnel can quantify the number of days required to conduct training. They can use this information to determine the number of ranges needed to support training requirements. For example, a mechanized infantry company must qualify its soldiers semiannually with the M16-series and M4 rifle. To meet this requirement, the unit conducts zeroing, practice, and day and night record firing. Two types of ranges support this firing:

(1) 25-meter zero range.

(2) 16-lane MRF range.

b. A mechanized infantry division includes five mechanized infantry battalions with four companies each. Based on single-company annual usage requirements of 3.4 days and 3.1 nights, compute one battalion's annual usage requirements as follows:

DAY: 3.4 DAYS PER COMPANY X 4 COMPANIES PER BATTALION = 13.6 OR 14 DAYS PER BATTALION

NIGHT: 3.1 NIGHTS PER COMPANY X 4 COMPANIES PER BATTALION = 12.4 OR 13 NIGHTS PER BATTALION

To determine battalion, and above, range throughput requirements, round up fractional days or hours to the next whole number.

c. Compute the total divisional annual M16-series and M4 rifle range requirement for mechanized battalions as follows:

DAY: 14 DAYS PER BATTALION X 5 BATTALIONS PER DIVISION = 70 DAYS PER DIVISION

NIGHT: 13 NIGHTS PER BATTALION X 5 BATTALIONS PER DIVISION = 65 NIGHTS PER DIVISION

d. An installation's M16-series and M4 rifle range requirement is determined by totaling the firing requirements of:

(1) Combat units (infantry, armor, field artillery, engineer, air defense, aviation, and cavalry).

(2) CS and CSS units.

(3) Installation support personnel.

(4) Reserve Component units.

(5) Reserve Officers' Training Corps units.

(6) Others requiring range support.

e. A comparison of this total with the range availability determines if the requirement can be met. If not, an availability shortfall exists that indicates the need for additional range capability.

f. Another example is a tank company must qualify its 14 crews semiannually on Gunnery Table VIII-A and B. Before firing for qualification, the unit practices on Table VII-A and B. Either the DMPRC or the DMPTR will support this training requirement.

g. Table 4-2 shows the approximate times for tank units to complete firing all offensive (Off) and defensive (Def) engagements on the tank tables shown. Time data include the time allowed for each engagement, five minutes between engagements, and ten minutes for the element to clear the range before the next element starts downrange. The time required to cycle a tank crew through a DMPRC or DMPTR is 74 minutes for Table VII-A and B and 76 minutes for Table VIII-A and B. Calculate the hours required for one company of 14 crews to train on the tables as follows:

**Table 4-2. Tank gunnery tables - time requirements (minutes)**

| Table | Crew | Section | Platoon |
|---|---|---|---|
| Table V-A and B | 95+ | | |
| Table VI-A and B | 104+ | | |
| Table VII-A and B | 74+ | | |
| Table VIII-A and B | 76+ | | |
| Table XII-A and B (Off and Def) | | | 120+ |
| + = Approximate times in minutes | | | |

TABLE VII-A and B: 74 MINUTES PER CREW X 14 CREWS PER COMPANY DIVIDED BY 60 MINUTES TO GET HOURS OR (74 x 14) ÷ 60 = 17.3 HOURS PER COMPANY.

TABLE VIII-A and B: 76 MINUTES PER CREW X 14 CREWS PER COMPANY DIVIDED BY 60 MINUTES OR (76 X 14) ÷ 60 = 17.7 HOURS PER COMPANY

h. Obtain the time requirement for a single battalion consisting of three tank companies plus the two headquarter's tanks to train semiannually is as follows:

TABLE VII-A and B: 17.3 HOURS PER COMPANY X 3 COMPANIES + 2 HQ TANKS PER BATTALION = 54.4 HOURS PER BATTALION

TABLE VIII-A and B: 17.7 HOURS PER COMPANY X 3 COMPANIES + 2 HQ TANKS PER BATTALION = 55.6 HOURS PER BATTALION

i. The annual requirement for the battalion is twice the semiannual requirement, or 104 hours for Table VII-A and B and 107 hours for Table VIII-A and B.

j. Obtain the total divisional usage requirements for five battalions:

TABLE VII-A and B: 104 HOURS PER BATTALION X 5 BATTALIONS PER DIVISION = 520 HOURS PER DIVISION

TABLE VIII-A and B: 107 HOURS PER BATTALION X 5 BATTALIONS PER DIVISION = 535 HOURS PER DIVISION

k. Based on an average of 20 hours per day available for training on DMPTRs and DMPRCs, the total division requirement is approximately 26 days for Table VII-A and B and 27 days for Table VIII-A and B:

TABLE VII-A and B: 520 HOURS PER DIVISION ÷ 20 HOURS PER DAY = 26 DAYS PER DIVISION

TABLE VIII-A and B: 535 HOURS PER DIVISION ÷ 20 HOURS PER DAY = 26.75 or 27 DAYS PER DIVISION

l. Range control personnel determine the remaining tank gunnery usage requirements (Tables IX, X, XI, and XII) and BFV gunnery requirements for units that will fire on the DMPRC or DMPTR. Once again, the projected number of available training days must be established and compared to the number of required range days. If the required days exceed the number of available days, a shortfall exists.

**4-7. Throughput Capacity**

a. Throughput refers to the total number of soldiers or units (e.g., crews, sections, squads, platoons, and companies) to be trained in a given period of time. The throughput capacity of a range is the number of soldiers or units that it can accommodate in a given period of time. For example, a rifle range might have a capacity of one company averaging 200 soldiers over a period of eight hours. This capacity would support at least one day-firing cycle and one night-firing cycle per day. Assuming the range is available 314 days per year, the range would have an annual throughput capacity of 314 companies, or 62,800 soldiers (314 companies x 200 soldiers per company), for day and night firing.

b. Calculation of throughput capacity is based on the type of training, the time required for a single individual or unit to complete a training event or series of events, and the period of time (day, week, month, year) which applies. Throughput calculations for some types of ranges may also include the number of soldiers or units that can train simultaneously.

c. For example, develop the daily throughput capacity of an MRF range with 16 lanes as follows:

NUMBER OF LANES: 16

TIME REQUIRED FOR ONE SOLDIER TO COMPLETE FIRING: 30 MINUTES OR 2 SOLDIERS PER HOUR

TIME AVAILABLE FOR TRAINING: 8 HOURS PER DAY

NUMBER OF SOLDIERS PER HOUR: 16 LANES X 2 SOLDIERS PER LANE PER HOUR = 32 SOLDIERS PER HOUR

THROUGHPUT CAPACITY = 32 SOLDIERS PER HOUR X 8 HOURS PER DAY = 256 SOLDIERS PER DAY

d. Obtain the annual capacity by multiplying the daily capacity by the number of days available for training:

DAILY THROUGHPUT CAPACITY = 256 SOLDIERS PER DAY

DAYS AVAILABLE FOR TRAINING: 314 (ESTIMATED)

ANNUAL THROUGHPUT CAPACITY = 256 SOLDIERS PER DAY X 314 DAYS PER YEAR = 80,384 SOLDIERS PER YEAR

## 4-8. Throughput Requirements

a. The number of required training iterations is included in the calculation to compute annual throughput requirements. For example, annual M16-series and M4 rifle training for infantrymen consists of:

(1) Two iterations of daytime zeroing with the weapon.

(2) Two day and two night iterations of qualification firing.

(3) Two day and two night iterations of sustainment firing.

b. Compute the total rifle training throughput requirement for a mechanized infantry division as follows:

NUMBER OF SOLDIERS IN COMPANY: 200

NUMBER OF COMPANIES: 30

NUMBER OF ITERATIONS REQUIRED: 6 DAY AND 4 NIGHT ITERATIONS OR 10 ITERATIONS PER YEAR

NUMBER OF SOLDIERS PER COMPANY PER YEAR = 200 SOLDIERS PER COMPANY PER ITERATION X 10 ITERATIONS PER YEAR = 2000 SOLDIERS PER COMPANY PER YEAR

THROUGHPUT REQUIREMENT = 2000 SOLDIERS PER COMPANY PER YEAR X 30 COMPANIES = 60,000 SOLDIERS PER YEAR

c. Comparing the throughput capacity to the throughput requirement (80,384 versus 60,000) reveals an excess capacity of 20,384 soldiers per year.

## 4-9. Armored Vehicle Gunnery Requirements

a. Tables 4-2 can be used to develop throughput requirements for firing gunnery Tables V through XII for tank battalions. Tables 4-3 and 4-4 can be used to develop throughput requirements for firing gunnery Tables V through XII mechanized battalions. Calculate the time required for a unit to fire a gunnery table by using the formula:

$H = N (M \div 60)$, where:

H is the number of hours required to complete the gunnery table.

N is the number of vehicle crews required to fire the table.

M is the number of minutes required for one crew to complete day and night firing events (divide by 60 to convert minutes to hours).

b. For example, the time required for a tank company consisting of 14 tank crews to complete Tank Table VIII-A and B (day and night) is:

N = 14 CREWS

M = 76 MINUTES (FROM TABLE 4-2)

H = 14 (76 ÷ 60) = 17.7 HOURS (17 HOURS AND 42 MINUTES)

c. Determine the number of days required for training by dividing the required hours by the number of hours that the range is available for daily training. In the example above if the Table VIII range were available 16 hours per day, the number of days required for a company to complete firing would be:

H ÷ 16 = 17.7 ÷ 16 = 1.1 DAYS (1 DAY, 1 HOUR, AND 36 MINUTES)

This can be expressed in a formula as:

D = N(M ÷ 60) ÷ A, where:

D is the number of days required to complete the table.

A is the number of hours that the range is available on a daily basis.

d. Throughput capacity for an MPRC or MPTR, where only one training element (tank) can complete a gunnery table at a time, is developed as follows:

NUMBER OF LANES: 1

TIME REQUIRED FOR ONE TRAINING ELEMENT TO COMPLETE TABLE: FROM TABLE 4-2

TIME AVAILABLE FOR TRAINING: 20 HOURS PER DAY (AVERAGE)

DAYS AVAILABLE FOR TRAINING: 277 (AVERAGE)

NUMBER OF TRAINING ELEMENTS PER DAY: NUMBER OF CREWS, SECTIONS, OR PLATOONS

THROUGHPUT CAPACITY = NUMBER OF LANES x ((TIME REQUIRED FOR TRAINING ELEMENT ÷ 60) x TIME AVAILABLE FOR TRAINING) x DAYS AVAILABLE FOR TRAINING

e. For example, a gunnery table requires an average of 87 minutes to complete. Therefore:

THROUGHPUT CAPACITY = 1((87 ÷ 60) X 20) X 277 = 8033 CREWS, SECTIONS, OR PLATOONS PER YEAR

f. Develop the tank gunnery training throughput requirement for an armored division assigned 15 tank companies with 14 crews per company as follows:

**For crews:**

NUMBER OF CREWS: 210 (15 TANK COMPANIES X 14 CREWS PER COMPANY)

NUMBER OF CREW TABLES: 4 TABLES PER CREW

NUMBER OF ITERATIONS: 2 PER TABLE PER YEAR

THROUGHPUT REQUIREMENT (TR) = 210 CREWS X 4 TABLES PER CREW X 2 PER TABLE PER YEAR = 1680 CREWS PER YEAR

**For sections:**

NUMBER OF SECTIONS: 105 SECTIONS (15 TANK COMPANIES X 7 SECTIONS PER COMPANY)

NUMBER OF SECTION TABLES: 2 TABLES PER SECTION

NUMBER OF ITERATIONS: 2 PER TABLE PER YEAR

TR = 105 SECTIONS X 2 TABLES PER SECTION X 2 PER TABLE PER YEAR = 420 SECTIONS PER YEAR

**For platoons:**

NUMBER OF PLATOONS: 60 (15 TANK COMPANIES X 4 PLATOONS PER COMPANY)

NUMBER OF PLATOON TABLES: 2 TABLES PER PLATOON

NUMBER OF ITERATIONS: 2 PER PLATOON PER YEAR

TR = 60 PLATOONS X 2 TABLES PER PLATOON X2 PER TABLE PER YEAR = 240 PLATOONS PER YEAR

**Total tank gunnery training:**

TOTAL TR = 1680 CREWS + 420 SECTIONS + 240 PLATOONS = 2,340 TRAINING ELEMENTS PER YEAR.

g. Comparing the throughput requirement to the capacity (2,340 versus 8,033) indicates an excess capacity of 5,693 training elements per year. This excess capacity can help meet BFV gunnery requirements of the division's mechanized battalions. Trainers develop the throughput requirement for these battalions in the same manner.

h. Table 4-3 displays the average time required to complete BFV Tables V-A and B through VIII-A and B and squad firing exercises. These times are not consistent worldwide. Trainers should adjust for local conditions including the level of range automation and the distance traveled down range between engagements.

**Table 4-3. BFV team/squad gunnery tables and exercises – time requirements (AP only) (minutes)**

| Table | Team/Squad |
|---|---|
| Table V-A and B | 68 |
| Table VI-A and B | 69 |
| Table VII-A and B | 81 |
| Table VIII-A and B | 81 |
| Infantry Squad Combat Exercise (Day and Night) | 120 |

i. Bradley Tables IX, X, XI and XII involve gunnery tasks associated with missions. Day and night engagements occur in one or more missions, as determined by the unit commander. Timelines for day and night engagements of these missions follow:

MOVEMENT TO CONTACT: 3.8 HOURS

ATTACK: 3.7 HOURS

RAID: 3.4 HOURS

AMBUSH: 3.3 HOURS

RECON/SECURITY: 3.6 HOURS

DEFENSE: 3.4 HOURS

RETROGRADE: 3.4 HOURS

AVERAGE: 3.5 HOURS

j. Table 4-4 displays the average time required for day and night firing on BFV Tables IX, X, XI, and XII.

**Table 4-4. BFV section/platoon gunnery tables – time requirements (hours)**

| Table | Section | Platoon |
|---|---|---|
| Table IX (Day and Night) | 3.5 | |
| Table X (Day and Night) | 3.5 | |
| Table XI (Day and Night) | | 3.5 |
| Table XII (Day and Night) | | 3.5 |

k. Scout sections perform Tables IX-A and B and X-A and B. Timing for these tables depends on terrain, weather, and distance between engagements. The estimated time for completion of these tables is eight hours for Table IX-A and B and eight hours for Table X-A and B.

### 4-10. Range Selection

Refer to Appendix A for the types of ranges needed to accomplish training events for selected weapon systems.

# Chapter 5
# Range Development

## 5-1. Introduction

This chapter provides procedures to implement the range modernization and standardization policies in AR 210-21. It discusses the range design and construction process and describes the quality assurance features in the range modernization program.

## 5-2. Planning

a. As identified in paragraph 1-5, the "range planners" coordinate the RTLP planning process in accordance with AR 210-21.

b. Planning an Army training range at installation level must be comprehensive, coordinated, and systematic. The first step is to determine facility needs based on the training and unit throughput requirements discussed in Chapter 4. The next step – range planning and location (the pre-design phase) – consists of project initiation, site analysis and selection, and documentation.

c. Planners must consider the resources available to support the development of training ranges. Project planning must follow the Army's planning, programming, budgeting, and execution system (PPBES) to ensure sufficient funds are allocated for construction and maintenance. Installation resource managers are familiar with the PPBES and can provide assistance to project planners.

d. The installation master plan must consider current and future use of installation and adjoining land when identifying suitable land for ranges and maneuver training areas. Maneuver areas should accommodate and be configured for units that will use them. Force structure, weapon systems, applicable doctrine, and METLs developed per 7-0 *Training the Force* and appropriate type unit Field Manuals. Training Circular (TC) 25-1, *Training Land,* is the guide to determining maneuver land requirements.

e. Planners evaluate requirements based on annual schedules, throughput capacity based on range or training land availability, and actual use. Requirements for additional range and maneuver area capabilities are valid where throughput requirements exceed the capacity of existing facilities.

f. Adverse environmental impact, safety and occupational health, and industrial hygiene aspects of range construction and operations must be taken into account. Contaminated impact areas will not be created unless an exception to policy has been approved per AR 200-1, *Environmental Protection and Enhancement,* and AR 210-20, *Master Planning for Army Installations*. Undertaking construction in an existing impact area will require expensive ordnance clearance to ensure the safety of construction personnel. Planned weapons impact areas must meet requirements for current weapon systems and those programmed for fielding. Planners should consider the effects of impact areas and SDZs on current ranges and maneuver areas; this avoids reducing the availability or capability of existing training areas. Planning should also address airspace requirements for combined arms training.

g. Planners must locate facilities and equipment storage to meet the needs of the proposed training and support planned land use. For example, grouping small-arms ranges with close-in training facilities reduces the need for separate support facilities (e.g., latrines, instruction buildings, and parking areas). Planners must also consider fuel consumption. Fuel is a critical resource so maneuver areas and ranges should be located as close as practical to cantonment and bivouac areas to minimize travel and provide expansion potential.

h. Sub-caliber ranges and TADSS should be used to conserve full caliber ammunition for zero, sustainment, and qualification firing.

i. Operations and maintenance plans for new facilities, including ranges and training areas must achieve manpower and other resource savings.

j. Documentation for new facilities should include information on the size, condition, and capacity of land, as well as the ranges and training facilities available locally or at other accessible training areas. Safety requirements, environmental considerations, and terrain profiles for proposed range locations should also be included. Much of this information is in the catalog of installation firing ranges required by the local range regulation. Land condition and capability data can be obtained from the installation directorate of public works (DPW) natural resources office. These data are based on land-condition trends determined by using the Integrated Training Area Management (ITAM) program. The ITAM program includes the geographic information system (GIS). Information on the ITAM program is available from Headquarters, Department of the Army (DAMO-TRS) .

k. Doctrine and training strategies, information on types of units, personnel, frequency, specific weapon systems, and assigned missions provide the basis for the range/training area planning phase. Use Standards in Training Commission (STRAC) guidance as a baseline to determine alternatives to construction or land acquisition.

l. Information on the arrival of a new weapon system and the associated facility requirements is contained in Army Modernization Reference Data (AMRD), distributed on compact disc – read only memory (CD-ROM) by HQDA (DAMO-FDR), and a support facility annex (SFA) published by the U.S. Army Corps of Engineers (USACE). An SFA-assessment of the facility impact of weapons development and fielding is accomplished through logistics management for major and directed acquisition programs per AR 700-127, *Integrated Logistical Support (ILS) Manager's Guide.* Support facility annex data are updated in coordination with the materiel system program executive officer (PEO) or program manager (PM) and TRADOC system manager (TSM) as weapon system development matures.

m. Engineering expertise is provided by:

(1) The divisional engineer staff officer.

(2) The local DPW or US Army National Guard facility management officer.

(3) The Corps of Engineers division or district engineer.

(4) The USACE RTLP Mandatory Center of Expertise (RTLP-MCX).

(5) The USACE Ordnance and Explosives Center of Expertise (OE-CX)

(6) COE laboratories.

n. The installation DPW includes training facilities requirements in the installation real property master plan.

o. The installation safety manager provides range safety expertise. This expertise is also available at each MACOM safety office. The technical advisor for range safety (TRADOC Safety Office) provides additional assistance. Special-use airspace matters are processed and coordinated under the provisions of AR 95-2, *Air Traffic Control, Airspace, Airfields, Flight Activities and Navigation Aids.* In addition to the installation air traffic and airspace officer and the DA regional representative (DARR) to the Federal Aviation Administration, policy and strategy guidance can be obtained from the U.S. Army Aeronautical Services Office, Fort Belvoir, VA.

p. Health and safety risk assessment and design review assistance may be obtained from the installation medical activity and safety offices, respectively. Additional assistance is available from:

(1) U.S. Army Center for Health Promotion and Preventive Medicine, ATTN: HPPM-MIL, Aberdeen Proving Ground, MD 21010-5422 (health assessment).

(2) U.S. Army Environmental Awareness Resource Center, Huntsville, AL (environmental support services and products).

(3) U.S. Army Environmental Center, ATTN: CETHA-TS-S, Edgewood Area, Building E4460, Aberdeen Proving Ground, MD 21010-5401 (design review assessment).

q. Military construction projects (both live-fire range construction and land acquisition) are long lead time requirements (five-to-seven years). Planners must consider practical alternatives. When evaluating alternatives or determining if existing training facilities can support new weapons systems or missions, planners should consider the:

(1) Number and types of training facilities available.

(2) Use rates.

(3) User density.

(4) Ammunition authorized.

(5) Serviceability or remaining life spans of facilities.

(6) Current and projected throughput requirements.

(7) Current use and potential for new or multiple uses.

(8) Adequacy to meet training standards.

(9) Life-cycle, operating, and maintenance costs.

(10) Adverse environmental impacts, including contamination of surface and ground water, bio-diversity, soil, cultural resources, deterioration of air quality and environmental noise impact.

(11) Contamination of impact areas and the potential for creating new contaminated impact areas.

(12) Land-use impact and compatibility, including current and future use of installation and adjoining land.

r. For additional assistance, contact:

(1) The MACOM.

(2) U.S. Army Training Support Center, ATTN: ATIC-ATMD, Fort Eustis, VA.

(3) U.S. Army Corps of Engineers, ATTN: CEMP-EA, Washington, DC.

(4) Range – Mandatory Center of Expertise (RNG-MCX), U.S. Army Engineering and Support Center, Huntsville, ATTN: CEHNC-IS-SP, Huntsville, AL.

(5) PEO-STRI, (Program Executive Office - Simulations, Training and Instrumentation Command (STRICOM)) Orlando, FL in accordance with AR 350-38.

**5-3. Project Initiation**

a. Project initiation includes:

(1) Identifying the training requirement.

(2) Identifying training facility needs based on adequacy and availability of existing facilities.

(3) Considering alternatives to facility modification or new construction.

(4) Developing a project milestone schedule.

(5) Collecting resource data.

b. Project initiation is a coordinated effort of planners from:

(1) The directorate of plans, training, and mobilization DPTM)/G3 (trainers, range officers, force developers).

(2) The DPW (facility engineers; installation master planners; environmental, natural, and cultural resource managers).

(3) Major using units.

(4) Installation resource management office.

(5) Security and safety agencies.

c. Representatives from the Directorate of Reserve Components, Staff Judge Advocate, and Public Affairs Office should participate as required.

d. Planners:

(1) Review training requirements.

(2) Consider alternatives to facility modification or new construction.

(3) Evaluate resources available to support development of a training range.

(4) Determine the specific type of range required, including weapon systems and ammunition.

(5) Consider construction impact on both the installation master plan and the RTLP.

**5-4. Range Land**

Land must be identified that accommodates the range layout, the SDZ of applicable weapon systems, and impact area requirements. An environmental analysis determines the impact of range construction and operations.

**5-5. Facilities and Equipment**

Resources must meet the needs of the proposed range and support nearby ranges. For example, collocating small-arms ranges with an obstacle or infiltration course would conserve land and resources by reducing common support facilities (e.g., latrines, instruction buildings, and parking areas).

**5-6. Training Schedules**

Range construction projects may impact training schedules for nearby ranges. Coordination with scheduling activities of the installation DPTM/G-3 must take place early in the planning process.

**5-7. Training Aids, Devices, Simulators, Simulations, and Sub-Caliber Ranges**

Use of these devices should be maximized to save full-caliber ammunition for zero, qualification, and sustainment firing. The acquisition of TADSS should be accomplished through AMC in accordance with AR 350-38.

**5-8. Resource Management**

a. Efficient use of manpower and resources for operations and maintenance should be planned.

b. Planners must consider the impact on operating costs when selecting a range location. Travel time and fuel are critical resources.

**5-9. Host Nations**

Range projects in overseas areas require coordination with host nation officials throughout the development process. Requirements will vary from nation to nation. The unit G-3 or local area commander should have information concerning any host nation agreement or legalities.

**5-10. Alternatives and Justification**

a. The Master Planning Board (MPB) approves decisions of installation planners to build new range facilities.

(1) Sharing use of an existing facility at another installation.

(2) Converting an older or specialized range to a multipurpose range.

(3) Using TADSS or additional portable targets on existing facilities.

(4) Increasing the hours of operation of existing ranges.

b. When deciding whether to share use of an existing range at another installation, commanders and planners consider:

(1) Availability of the other range.

(2) Distance and cost of travel to the alternate range.

(3) Frequency of firing required to meet weapons training standards.

(4) Possible limitations on the use of ammunition required for training.

(5) Obsolescence of equipment, facilities, or range design.

(6) Potential restrictions and waivers on range use.

c. Factors bearing on justification of a facility include:

(1) The overall range facility needs to meet valid training requirements and throughput.

(2) Fielding needs to meet unit training requirements and throughput levels.

(3) A change in the installation mission or force structure requiring an increased capability to conduct training.

(4) Capability of standard ranges to meet homestation peacetime and mobilization training requirements. For example, modernization of one range could satisfy training needs and allow the installation to close other ranges to free land for other training needs.

d. The construction and use of multipurpose training ranges ensure better management of training activities, facilities, and land.

**5-11. Site Planning Process**

a. Once the decision to develop a range is made and approved according to AR 210-21, the site planning process begins. The steps in the procedure are:

(1) Analyze training task requirements.

(2) Determine the type of range needed.

(3) Identify candidate sites.

(4) Analyze sites.

(5) Select the best site.

b. Site planning is crucial to successful training facility planning and development. Site planning must comply with requirements outlined in AR 200-1, AR 200-2, *Environmental Effects of Army Actions*, AR 210-20, TC 25-1, and this training circular.

c. The trainer works with the installation DPW (including environmental, natural, and cultural resource managers) and safety officials to coordinate range-site decisions that enable effective training. A new or modified range may have environmental impact, especially noise, on the installation and nearby communities. Planners should minimize the impact, but they must maintain a balance between environmental concerns and training requirements. For example, keeping maximum natural vegetation promotes realism and reduces environmental damage. Information on planting programs that enhance cover and concealment for training realism is available from the U.S. Army Construction Engineering Research Laboratory's pamphlet, *Tactical Concealment Area*, August 1999.

d. Planners consider the location of the proposed range in relation to the rest of the installation. The site should not isolate useful maneuver terrain, cut off impact areas, or make clearance operations difficult. When the proposed range requires frequent maintenance, the site should provide easy access when adjacent ranges are active. Unusable terrain should be incorporated into impact areas. Moreover, the range should have potential for growth and improvement to accommodate new weapon systems. The key to efficient operations is a site that is easily accessible to using units. Planners need to consider overall training requirements and the flow to and from ranges when determining site locations. Training areas, such as small-arms ranges and bayonet courses, should be close to garrisons. Tracked-vehicle and artillery ranges may be located further away from unit housing. When ranges are remote, planners should consider overnight parking and bivouac facilities.

e. Due to the shortage of lands for live-fire training, ranges are sited over existing ranges or extended to encroach on existing impact areas. The trainer must consider the increased financial impact and construction time required to clear the site of unexploded ordnance to ensure safe construction operations.

**5-12. Terrain**

a. Terrain configuration will influence the extent of construction and the cost required to transform land into a usable range. Contour profile, soil content, and trafficability requirements affect the extent of moving and other clearing work needed. Sites requiring minimal earthwork to create level trails, visible targets, and good drainage are preferred.

b. Terrain has a major effect on direct-fire weapon systems. It influences target locations, target visibility, and a soldier's ability to fire on moving targets. Training in challenging terrain with well-planned scenarios can improve a soldier's ability to hit a moving target. Terrain features that influence target location are:

(1) Topography.

(2) Soil characteristics.

(3) Surface composition.

(4) Hydrology.

(5) Vegetation.

(6) Man-made obstacles.

c. These features also directly affect range maintenance and the environmental impact of construction and operations.

d. A terrain profile of the proposed range should be developed to support feasibility estimates, review safety requirements, and determine target locations.

e. Range sites must be selected to ensure that terrain allows maximum use of weapon systems. Planners attempt to develop ranges in areas with different types of terrain to promote weapon-firing skills under varying conditions. The site should slope downward from the firing points to the target area. This assists drainage, provides visibility, and helps contain projectiles within the SDZs. Fire-and-movement ranges should have both natural and manmade cover and concealment islands. Vegetation, however, should be removed only to provide a clear view of targets. Planners must balance the need for realism with unit training objectives. A land maintenance plan provides for regrading, revegetation, and erosion control to maintain training realism and reduce long-term maintenance costs.

f. Water areas contribute to training and increase combat realism. Reflections and temperature variations over or near water create conditions that can only be learned through experience. However, unexploded (dud) ordnance embedded in shallow water could hamper future land clearing and decontamination operations. Precautions must also be exercised if lasers are used on a range; the reflection of laser beams on water is a potential eye hazard, and open water directly in front of or behind targets could render the range unusable with laser devices. Water areas should be sited as part of the overall storm water runoff and erosion control without detrimentally affecting the use of the range.

**5-13. Range Impact Area Requirements**

Impact area requirements vary with training requirements and with SDZs. New contaminated impact areas will not be created unless an exception to policy is approved per ARs 200-1, 210-20, and 210-21. Conveniently located hill masses or other natural terrain features may significantly reduce real estate requirements for SDZs. Ranges should be sited so they do not restrict maneuver areas. Refer to AR 385-63, *Range Safety*, and DA PAM 385-63, *Range Safety*. Impact area requirements are based on:

a. The number and type of ranges required.

b. The SDZ for each type of weapon.

c. The use of common impact areas, if possible.

**5-14. Gun-Target Lines**

a. The gun-target line (GTL), also known as the "line of fire," is an imaginary line drawn from a weapon system (firing position) to a target. For efficient small-arms range operations, it is best to orient the GTL to the northeast in the Northern Hemisphere. This places light on the face of the target for the longest part of the day. Orient small-arms marksmanship ranges so that soldiers are not required to fire into the rising or setting sun. However, direction of fire is not as important as accessibility, security, and safety.

b. Consider terrain configuration when orienting the GTL in order to reduce the area covered by the SDZ. If a site is on rough terrain, the GTL should be perpendicular to high ground. Since small-arms ranges are usually on flat terrain, the GTL should be horizontal or slightly below horizontal. Range planners must avoid sites requiring the GTL for direct-fire, line-of-site weapon systems to pass over public roads or navigable waters. If firing over installation roads is unavoidable, it is mandatory that these firing areas be marked with appropriate signs and separate warnings and closed during firing periods. Indirect artillery firing over public access areas must comply with procedures in AR 385-63 and DA Pam 385-63.

c. Planning GTLs for vehicle-mounted weapon systems presents a greater challenge. Vehicular weapon systems such as tanks, Bradleys, and Strykers require that GTLs be planned for multiple targets from stationary firing points and for single and multiple targets from a moving tank or Bradley Fighting Vehicle.

d. Firing on the move requires planners to analyze a series of GTLs from various firing points along the path of the moving vehicle or within designated "firing boxes." Multiple firing scenarios for combat vehicle ranges that include both stationary and moving gun platforms and stationary and moving targets require a series of GTLs with their respective SDZs superimposed over common terrain.

e. Set safety requirements for moving vehicle firing exercises by designing a course with sequential firing points perpendicular to the target area. This arrangement may require several kilometers to accommodate training with high-speed vehicles.

f. The result is a range that needs more depth than one with the vehicle firing from a fixed point. A course on which a vehicle is required to change direction frequently more closely represents tactical conditions. It also increases land requirements since it requires larger SDZs. Refer to AR/DA Pam 385-63 for procedural guidance on this type of GTL and SDZ planning.

**5-15. SDZ Requirements**

a. In the most economical layout for a range complex the impact areas of SDZs overlap, but the maneuver or firing areas themselves do not. This reduces the overall acreage needed for the range system, which conserves land and minimizes the size of impact areas that are contaminated.

b. Some ranges should be sited close together, such as 25-meter zero ranges and record-fire ranges. Some ranges are incompatible for contiguous siting. For example, field artillery firing points should not be located next to rifle ranges since the field artillery noise detracts from rifle marksmanship training.

## 5-16. Target Areas

a. Once the range layout is approved, planners must design and develop the target area. Target areas on ranges used for individual and collective training should provide realistic training and meet gunnery qualification requirements.

b. Target arrays must:

(1) Incorporate valid training requirements.

(2) Meet size, quantity, and distance requirements of gunnery manuals.

(3) Make logical use of terrain for target arrangement.

(4) Provide a combination of fire and maneuver when necessary to meet training requirements.

## 5-17. Target Visibility

Firing points and targets must be located to maximize flexibility for scenario development and to meet training criteria. Planners must position targets within safe limits and be sure they are visible from firing points, positions, or lanes. To meet training requirements, 90 percent of the target must be visible from the firing position. Simulation of combat situations is important. When threat forces are portrayed on a range, groups of targets that may be visible to personnel in several firing positions will appear at the same time. If visual obstacles are used, their placement should allow reasonable target exposure time. The key is to provide an environment in which correct gunnery techniques can be exercised using realistic battlefield obstacles and target speeds.

## 5-18. Range Analysis and Design System

a. The Range Analysis and Design System (RADS) provides a graphical and analytical profile from each firing position to each target on a range. This stored data can be sorted to produce a target catalog. This catalog displays the engagement pairs and distance between them, meeting the criteria for the training requirement found in armor, Bradley and Stryker gunnery FMs. A line of sight (LOS) analysis by RADS should be performed concurrently with the concept design review of the range design and again at the final review. This process ensures that the targets are visible to meet training requirements. RADS assists the master gunners or equivalent in the initial layout of a range and the development of training scenarios. The LOS analysis is required for all ranges, if possible. The design engineer normally has the capability to accomplish this function.

b. The LOS analysis is performed by the U.S. Army Engineering and Support Center, Huntsville (USAESCH). Approximately 30 days is required to complete the analysis and a master gunner is required in Huntsville for approximately one week to assist with the analysis. The following data are required to perform an LOS analysis:

(1) Digital 3D topographical data of the range site.

(2) Proposed layout of the range (including digital design data, XY Z coordinates).

### 5-19. Target Accessibility and Protection

a. Complex target-elevating mechanisms are maintained most economically in central maintenance facilities equipped for the task. Range maintenance trails are needed for access to target-elevating mechanisms so that they can be removed from the range for maintenance.

b. Targets should be placed at distances from firing points that match training guidance, taking advantage of visibility, cover, and concealment. Landmarks at the target location, such as tree lines, manmade objects, or streams and roads, help soldiers estimate distance to a target. Gunnery manuals, this training circular, and (USAESCH) CEHNC1110-23 series standard design manuals provide guidance and distance requirements for target placement by the type of range.

c. Stationary target-elevating mechanisms and moving target carriers need protection from damage. Target emplacements prepared during construction protect target mechanisms. Natural terrain can provide additional protection and reduces the extent of the earth-moving operations needed to construct a large protective berm.

### 5-20. Gun-Target Engagement Sequence

Develop and plot gun-target engagement sequence capability to ensure that weapons can be contained within the designated impact area without endangering personnel or creating a hazard to property or facilities. Although primarily used on multi-purpose range complexes, a gun-target engagement sequence matrix is a valuable tool that helps determine target array options on any range. The matrix lists all targets that can be engaged from each firing position and the distance to each target. It identifies the primary targets for each scenario.

### 5-21. Preliminary Documentation

a. After site selection, the range-planning team prepares a preliminary range site layout. The layout shows a draft diagram (rough sketch format) of the site outline, downrange area, firing line and maneuver boxes, and support facilities. The team prepares perimeter descriptions and layout maps following guidance in AR 210-20, AR 415-15, *Army Military Construction Program Development and Execution,* and AR 415-10. Perimeter descriptions and layout maps show the proposed range area with existing and planned facilities. A site description may also be needed for environmental impact assessments and statements per AR 200-1 and AR 200-2. The planning team coordinates requirements incorporated into the development project, such as construction, equipment, facilities, and safety, with the installation:

(1) Director of Public Works.

(2) Safety Officer.

(3) Public Affairs Officer.

(4) Other appropriate staff offices.

b. Perimeter descriptions and site maps portray the proposed training area with existing and planned facilities. Planners assist in preparing the project summary. A preliminary construction layout is included in the project summary to show the proposed training area in relation to troop housing areas, road networks, existing ranges, and maneuver areas. Layouts will use current mapping data and may be supplemented with aerial photographs.

**5-22. Project Summary**

a. Once the target area is designed and developed, it is added to the preliminary layout and project summary. Trainers assist engineers to prepare the project summary. It includes a preliminary construction layout to show the proposed range in relation to troop housing areas, road networks, existing ranges, and maneuver areas.

b. The project summary includes general information concerning design and construction. Materials specified for use must conform to local construction policies. Grading, normally held to a minimum, is used to provide adequate drainage, increase target visibility, prevent ricochets, and meet other engineering requirements. A program for seeding, sod and other soil erosion control measures must be established to reduce maintenance costs and enhance operations. Compaction requirements such as embankment, drainage, and general cleanup must meet Corps of Engineers standards. The next step is approval of the project summary by the installation Master Planning Board (MPB).

# Chapter 6
# Range Operation

## 6-1. Introduction

a. Installation or community commanders will establish range-control and safety programs according to AR 210-21 and AR 385-63. They will appoint range-control personnel to supervise weapons firing on the installation and enforce safety and operational requirements. The safety manager provides oversight responsibility and monitors the effectiveness of the installation range safety program.

b. This chapter describes the range-control organization and its responsibilities to schedule and maintain positive control over range activities. It describes procedures to follow during live-fire and lasing operations. Refer to Appendix C for a checklist to assist these operations.

## 6-2. The Range Organization

a. The installation or community range management function is normally found in the Directorate for Plans, Training, and Mobilization (DPTM) or G-3. If the installation range and training functions were consolidated under a single division, it would be titled the "Range, Training, and Training Support Division"; however, the functions are typically assigned to separate training and range divisions. If a range division is not warranted, a training support division performs range functions and is the point of contact between the installation and the U.S. Army Training Support Center (USATSC).

b. The range division/training support division plans, controls, and supervises weapons firing on the installation. Duties of a range division include action to:

(1) Direct and enforce risk management controls pertaining to firing on ranges.

(2) Plan, schedule, and operate the installation's ranges and perform organizational maintenance.

(3) Plan, schedule, supervise use of facilities, training areas (including restricted airspace), and supporting weapons.

(4) Develop, publish, maintain, and enforce range and training area regulations, standing operating procedures (SOP), airspace procedures, and firing bulletins.

(5) Requisition, store, and issue targets, target materials, and equipment.

(6) Ensure that ranges are maintained and capable of meeting training needs in accordance with the Army's ITAM Program. The ITAM program establishes procedures to achieve optimum, sustainable use of training ranges and lands by implementing a uniform land management program.

c. The DPTM or G-3 training division plans, coordinates, and supervises the installation military training program. Many duties relate directly to the range program to include:

(1) Implement and supervise training programs that increase training effectiveness through substitution, simulation, and miniaturization.

(2) Plan and coordinate implementation of competitive marksmanship programs.

(3) Implement, supervise, and evaluate soldier and collective training for installation units.

(4) Supervise and manage battlefield simulation exercises.

(5) Prepare the installation range development plan.

d. Another DPTM or G-3 division with duties relating to the range program is the mobilization division. This division plans for range and training area requirements during mobilization. This division also assists the range division to develop the installation's land use requirements study (LURS) for post-mobilization training requirements (see TC 25-1).

**6-3. Range Control and Scheduling**

a. Coordination and integration of installation training and firing activities requires effective range control and scheduling procedures. Communications augment these procedures to ensure safe, efficient live-firing activities. Range control begins with development and publication of the installation range regulation. This regulation establishes policies and procedures for range activities consistent with safety and environmental regulations. Installations supplement the safety and environmental regulations with SOPs applicable to individual facilities within the installation-training complex. AR 210-21 describes the range regulation subjects and the minimum content requirements for an SOP.

b. An important feature of a range-control program is an effective range-scheduling process. Scheduling decisions consider range-throughput capacities, usage rates, applicable limitations or restrictions, special training needs identified by unit commanders, and training priorities. A critical scheduling consideration is the elimination of safety conflicts between ranges and maneuver areas.

c. Range Facility Management Support System (RFMSS) assists range-management operations and automates the scheduling process. The database maintains useful data on ranges, training areas, ammunition, range conflicts, and using units. Most Army installations, including RC sites, use RFMSS.

(1) RFMSS is an automated system that will simplify and enhance the current process of range and training facility management and scheduling functions at an installation. It will provide the capability for a unit to electronically submit requests for reservations. Authorized users will be able to view available and scheduled facilities for approval, disapproval, and rescheduling of results; create two and three dimensional SDZs; and detect scheduling, environmental, and safety conflicts. RFMSS will track, collect, monitor, and display range specific information.

(2) Remote access to RFMSS permits unit training personnel to identify available resources and submit schedule requests.

(3) Range-control personnel review and approve these requests and ensure that required information has been provided and ensure scheduling, safety, or environmental conflicts are resolved.

(4) The RFMSS produces a variety of useful reports such as daily and weekly scheduling bulletins, range-facility utilization reports, range-cancellation and no-show reports, ammunition-expenditure and ADHOC reports. RFMSS also uses a Geographic Information System to provide a graphic fire desk and SDZs as overlays to range maps.

d. An effective communications system is essential to range operations. Range Control is the Network Control Station and any orders it issues, such as a general cease fire order must be obeyed immediately. Range officials use the communications system to control firing, coordinate requests for medical assistance, and announce unsafe conditions. Two-way communications must be maintained between the range control organization and using units during firing and weapons training periods. If communications are disrupted, the unit must cease firing and suspend operations until they are reestablished. A range-control communications network normally consists of:

(1) A frequency-modulated (FM) radio range-control firing net.

(2) An FM radio range-control administrative net.

(3) A radio or telephone backup range-control firing net.

e. The network provides communications between the range-control network and each firing range, weapons training facility, and training area within the installation-training complex. The primary communications system may be either FM radio or telephone. The using unit must report firing position openings, closings, and changes in firing status immediately to the range-control communications center. If the RFMSS is used, a communications program is available to record event data. The range-control communications operator can enter training data into the RFMSS database, such as range openings, closings, and firings and, if necessary, print out to provide a detailed reconstruction of all calls and events.

f. Plots of SDZs (described in DA Pam 385-63) play an important part in range scheduling. They are the primary determinant of potential safety conflicts between ranges. Vertical danger zones should also be computed and considered to accommodate the ballistic trajectory and ricochet characteristics of weapons and ammunition that will be used. The scheduling process should also incorporate checks to ensure that two or more units are not scheduled on the same firing range at the same time without a co-use agreement between the commanders concerned. Periodic scheduling conferences may assign major training activities to available training facilities and reconcile non-training requests for range use with the range schedule.

**6-4. Live-Fire Operations**

a. Safety is paramount during live-fire and laser operations. Using-unit commanders (battalion or squadron) will ensure that the officer-in-charge (OIC), the range safety officer (RSO), and the laser range safety officer (LRSO) receive an installation or community safety instruction before obtaining battalion or squadron commander safety certification. The safety certification programs will train and qualify personnel in OIC, RSO, and LRSO duties for firing exercises or maneuver operations. AR 385-63 and DA PAM 385-63 are the primary source for range safety responsibilities.

b. The OIC is responsible for overall safety of training and proper use of the training facility. Range operations and personnel requirements will be in accordance with AR 210-21, AR 385-63, DA Pam 385-63 and local safety SOP. OIC qualifications and duties are set forth in DA PAM 385-63.

c. The RSO/LRSO will be a commissioned officer, warrant officer, NCO, or a civilian in the rank/grade established in DA PAM 385-63. The RSO/LSRO makes the final determination to fire. The RSO/LSRO must be weapon-system-qualified and have completed a program of instruction in RSO/LSRO duties. The RSO/LSRO must be present on the training complex and have no other duties that period of training, except for aviation weapon systems where instructor pilots may assume RSO/LSRO duties. Assistant range safety officers (ARSO) may be appointed as required. RSO/LSRO qualifications and duties are set forth in DA PAM 385-63.

**Appendix A**
**Range Selection**

**This appendix identifies primary and alternate ranges used for training and qualification with specific weapon systems and weapons, based on applicable field and gunnery manuals. Table A-1 lists primary and alternate ranges to satisfy weapon-system training events. The Army's Stryker Brigade Combat Team (SBCT) will require ranges to train soldiers and crews. Table A-2 reflects the estimated SBCT training support infrastructure requirements and ranges to support training.**

**Table A- 1. Range selection**

| WEAPON SYSTEM | TRAINING EVENT | RANGE | |
|---|---|---|---|
| | | PRIMARY | ALTERNATE |
| **M16-Series and M4** | Zero | 25m Range | AFF/ARF/MRF with 25m boots |
| | Sustainment/Record | AFF/ARF/MRF | 25m Range |
| | Qualification | ARF/MRF | 25m Range |
| | Night | AFF/ARF/MRF Night | 25m Range |
| | NBC Conditions | AFF/ARF/MRF | 25m Range |
| | Squad/Platoon ARTEP LFX | ISBC/IPBC/ MPTR/MPRC | MPTR/MPRC |
| **9mm** | Instructional Fire | CPQC | 25m Alternate Pistol Qualification Course |
| | Qualification | CPQC | 25m Alternate Pistol Qualification Course |
| | NBC Conditions | CPQC | 25m Alternate Pistol Qualification Course |
| | Military Police Qualification | MPFQC/CPQC | 25m Alternate Pistol Qualification Course |
| | Military Police Night Fire Arms Sustainment | MPFQC/CPQC | 25m Alternate Pistol Qualification Course |
| **M21 and M24** | Zero | Sniper Field Fire | None |
| | Qualification | Sniper Field Fire | None |
| | Squad/Platoon ARTEP LFX | Sniper Field Fire | MPTR/MPRC |
| | CALFEX/STX | MPRC | None |
| | Night Fire | Sniper Field Fire | None |

**Table A-1. Range selection (cont)**

| WEAPON SYSTEM | TRAINING EVENT | RANGE PRIMARY | RANGE ALTERNATE |
|---|---|---|---|
| **M249 SAW/M240B** | 10m Practice<br>10m Record<br>Transition Practice<br>Transition Record<br>NBC Conditions<br>Squad/Platoon ARTEP<br>LFX/STX<br>CALFEX | Machine Gun 10m<br>MPMG<br>MPMG<br>MPMG<br>MPMG<br>ISBC/IBPC/<br>MPTR/MPRC<br>MPRC | None<br>None<br>None<br>None<br>None<br>None<br><br>None |
| **M60/M2** | 10m Record<br>Transition Record<br>Limited Visibility/<br>Predetermined Fire<br>Squad/Platoon ARTEP<br>LFX<br>CALFEX | Machine Gun 10m<br>MPMG<br>MPMG<br><br>ISBC/IPBC/<br>MPTR/MPRC<br>MPRC | None<br>None<br>None<br><br>None<br><br>None |
| **M203/M79** | Instructional<br>Qualification<br>NBC<br>Squad/Platoon ARTEP<br>LFX<br>CALFEX | Grenade Launcher<br>Grenade Launcher<br>Grenade Launcher<br>ISBC/IPBC<br>MPTR/MPRC<br>MPRC | None<br>None<br>None<br>None<br><br>None |
| **Hand Grenade** | Instructional | HG Qualification Course | None |
| | Qualification | HG Qualification Course | None |
| | Live-Grenade Throw | HG Live Fire Range | None |
| **MG/MK-19/Sniper Wpns** | Practice | MPMG/Sniper FF | None |
| | Record Fire | MPMG/Sniper FF | None |
| | NBC Condition | MPMG/Sniper FF | None |
| | Night Firing | MPMG/Sniper FF | None |
| | HE | MPMG/Sniper FF | None |
| | MG Sustainment | MPMG/Sniper FF | None |
| | LFX | MPMG/Sniper FF | None |
| | CALFEX MPRC | MPMG/Sniper FF | None |
| **60mm Mortar** | Crew/Section/Platoon Training | Mortar | None |
| | Evaluated ARTEP | Mortar | None |
| | CALFEX | Mortar | None |
| **81mm Mortar** | Crew/Section/Platoon Training | Mortar | SMR* |
| | Crew/Section/Platoon Training | Mortar | None |
| | Evaluated ARTEP | Mortar | None |
| | CALFEX | Mortar | None |
| **? Scaled mortar range** | | | |

**Table A-1. Range selection (cont)**

| WEAPON SYSTEM | TRAINING EVENT | RANGE PRIMARY | RANGE ALTERNATE |
|---|---|---|---|
| **120mm Mortar** | Crew/Section/Platoon Training | Mortar | SMR |
| | External Evaluation | Mortar | None |
| | CALFEX | MPRC | None |
| **M1-Series** | Table I | UCOFT* | None |
| | Table II | UCOFT* | None |
| | Table III | SGR*** | Maneuver Area Day-fire |
| | Table IV** | MPTR/DMPTR/ | MPRC/DMPRC |
| | | MPTR/DMPTR | |
| | Table V | | MPRC/DMPRC |
| | Calibration | Screening/Zero Range | None |
| | Table VI | MPTR/DMPTR | MPTR/MPRC |
| | Table VII | MPTR/DMPTR | MPRC/DMPRC |
| | Table VIII | MPRC/DMPRC | MPTR/DMPTR |
| | Table IX | MPRC/DMPRC | MPTR/DMPTR |
| | Table X | MPRC/DMPRC | MPTR/DMPTR |
| | Table XI | MPRC/DMPRC | None |
| | Table XII | MPRC/DMPRC | None |

* Unit Conduct of Fire Trainer
** Tank crew proficiency course (TCPC)
*** Scaled gunnery range
**** Tactical engagement simulation

| WEAPON SYSTEM | TRAINING EVENT | RANGE PRIMARY | RANGE ALTERNATE |
|---|---|---|---|
| **Bradley Fighting Vehicle (Linebacker and Bradley Engineer Vehicle** | Tables I-IV | MPTR/DMPTR | MPRC/DMPRC |
| | Calibration | Screening/Zero Range | None |
| | Table V | MPTR/DMPTR | MPRC/DMPRC |
| | Table VI-A/B | MPTR/DMPTR | MPRC/DMPRC |
| | Table VII-A/B | MPTR/DMPTR | MPRC/DMPRC |
| | Table VIII-A/B | MPRC/DMPRC | MPTR/DMPTR |
| | Infantry Squad Combat Exercise | MPRC/DMPRC | MPTR/DMPTR |
| | Table IX* | MPRC/DMPRC | MPTR/DMPTR |
| | Table IX-A/B** | MPRC/DMPRC | MPTR/DMPTR |
| | Table X-A/B** | MPRC/DMPRC | MPTR/DMPTR |
| | Table XI | MPRC/DMPRC | MPTR/DMPTR |
| | Table XII | MPRC/DMPRC | None |
| | CALFEX | MPRC/DMPRC | None |

* Vehicle section gunnery

** Scout section gunnery

**There must be installation-level training infrastructure to support the Army's transformation initiative. The Stryker Brigade Combat Team installation training requirements and ranges are shown in Table A-2. These requirements were developed by Training and Doctrine Command and accepted by Headquarters Department of the Army through the Requirements Review Prioritization Board process. The requirements were also adopted for the Assistant Chief of Staff for Installation Management template to identify infrastructure requirements for Transformation.**

## Table A-2. Estimated Stryker Brigade Combat Team training requirements and ranges

| TRAINING ASSET REQUIREMENTS |
|---|
| Simulator building (motion-based). |
| Simulator building (non-motion based). |
| Moving target simulator building. |
| Simulations center. |
| Maneuver/training area (complex terrain) - battalion 600 $KM^2$ and brigade 2500 $KM^2$. |
| Robust inter- and intra-installation C4I network (which includes a fixed tactical internet capability). |
| Integrated automation tools for development, storage, distribution, and management of training products. |
| Sufficient Army Distance Learning Program equipped classrooms for institutional and self-development training support. |
| MILES 2000 and future One TES. |
| Home station instrumentation system. |
| TWGSS/PGS-like system for gunnery training. |
| Engagement Skills Trainer. |
| Conduct of fire or artillery fire integration support team gunnery trainer. |
| CCTT-like system for the Interim Armored Vehicle. |

| NUMBER | RANGES | |
|---|---|---|
| | PRIMARY | ALTERNATE |
| 1 | 25m Zero Range | None |
| 1 | Modified Record-Fire Range | None |
| 1 | Combat Pistol Qualification Course | 25m Alternate Pistol Course |
| 1 | Multipurpose Machine Gun & Sniper Field Fire | None |
| 1 | Grenade Launcher Range | None |
| 1 | Sniper Field Fire Range | SPMG/MPTR/MPRC |
| 1 | MK-19 Range (for Anti-Tank (AT)-4, Javelin training also) | None |
| 1 | Hand Grenade Qualification Course | None |
| 1 | Anti-Armor Tracking Range | MPTR/DMPTR |
| 1 | Mortar | None |
| 1 | Multipurpose Training Range | None |
| 1 | Aerial Harmonization Range | None |
| 1 | Aerial Gunnery Range | MPRC |
| 1 | Digital Multipurpose Training Range | BAX |
| 1 | Battle Area Complex (BAX) | Digital Multipurpose Range Complex (DMPRC) |
| 1 | Infantry Squad Battle Course | None |
| 1 | Infantry Platoon Battle Course | None |
| 1 | Urban Assault Course (contains Live Fire Breach Facility and Live Fire Shoot House) | None |
| 1 | Combined Arms Collective Training Facility | None |

## Appendix B
## Range Equipment

Figures B-1 through B-69 illustrate targets and targetry equipment used to support ranges. For additional information, contact USATSC ATTN: ATIC-ATMD, Fort Eustis, Virginia 23604-5166, DSN 827-2320.

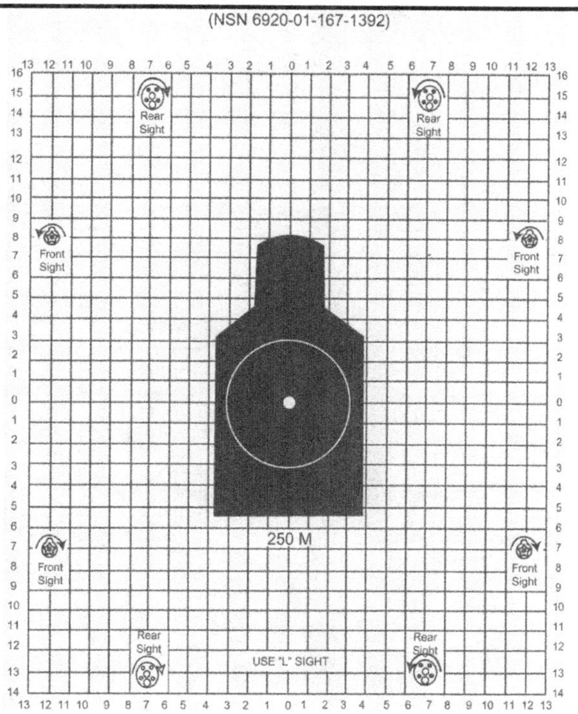


1. **Aim at target center. Adjust sights to move shot group center as close as possible to white dot.**

2. **At completion of zero, rotate rear sight to unmarked aperture and weapon will be battlesight zero for 250 meters.**

**Figure B-1. 25-meter zeroing target for M16A1 rifle (standard sights)**

**NOTES:**

1. **Aim at target center. Adjust sights to move shot group center as close as possible to white dot.**
2. **Use unmarked aperture with standard sights.**
3. **Use "L" sight with low light level sight system.**

**Figure B-2. 15-meter qualification zero target**

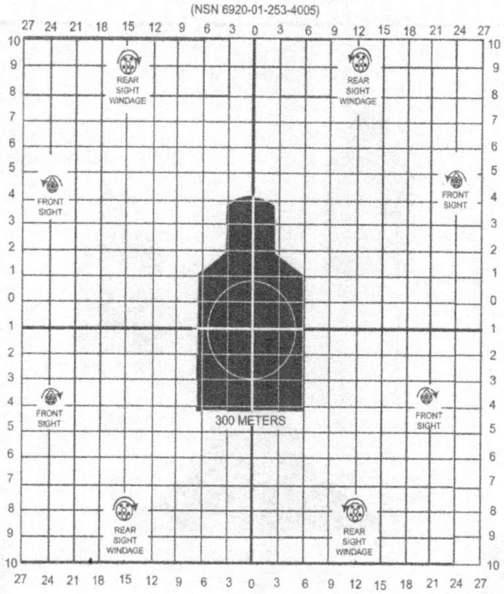


**ZERO TARGET DATA:**

1. **Rotate rear sight elevation knob to the 8/3 setting then up (right) one click past the 300 mark for zeroing at 25 meters.**

2. **Aim at target center. Adjust sights b move shot group center as close as possible to the center of the target.**

3. **After completion of the 25-meter zero rotate the rear sight elevation knob back one click to the 300/800 meter mark. The weapon will be zeroed for 300 meters.**

**Figure B-3. 25-meter zeroing target M16A2**

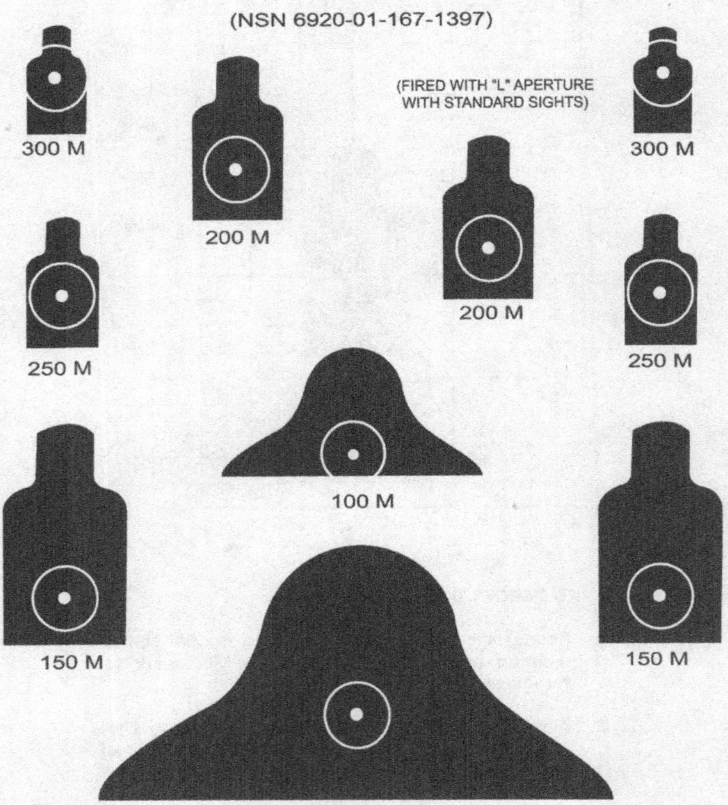


NOTE:

**Use long-range sight. The white dot on each target shows the best aiming point for targets at actual distances. If adjusted aiming point is used at 25 meters, bullets should hit within the circles, but are scored as hits if they hit anywhere in the silhouette.**

**Figure B-4. 25-meter scaled silhouette timed-fire target**

(NSN 6920-01-167-1391)

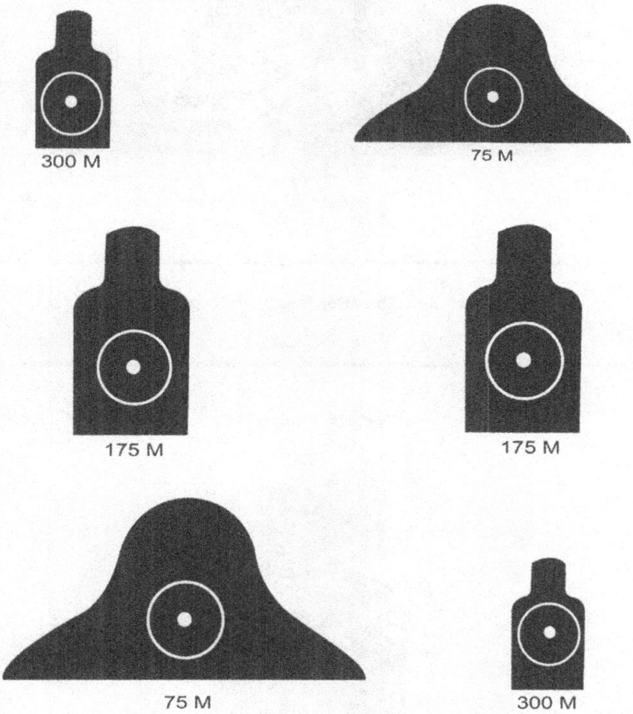


**NOTE:**
**Use long-range sight. The white dot on each target shows the center-of-mass aiming point. Bullets should hit within the circles, but are scored as hits if they hit any part of the silhouette.**

**Figure B-5. 25-meter scaled silhouette slow-fire target**

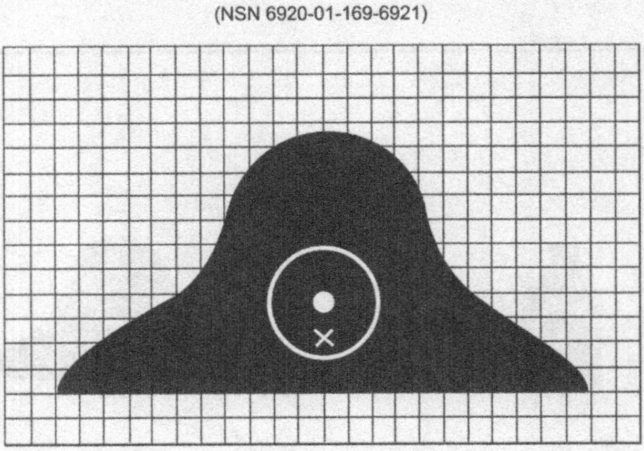


Figure B-6. 75-meter feedback target

**NOTE:**

**Aim at the X, move shot group to center, and click each sight the number of times marked on line. Allow for wind before making sight adjustments.**

Figure B-7. 175-meter feedback target

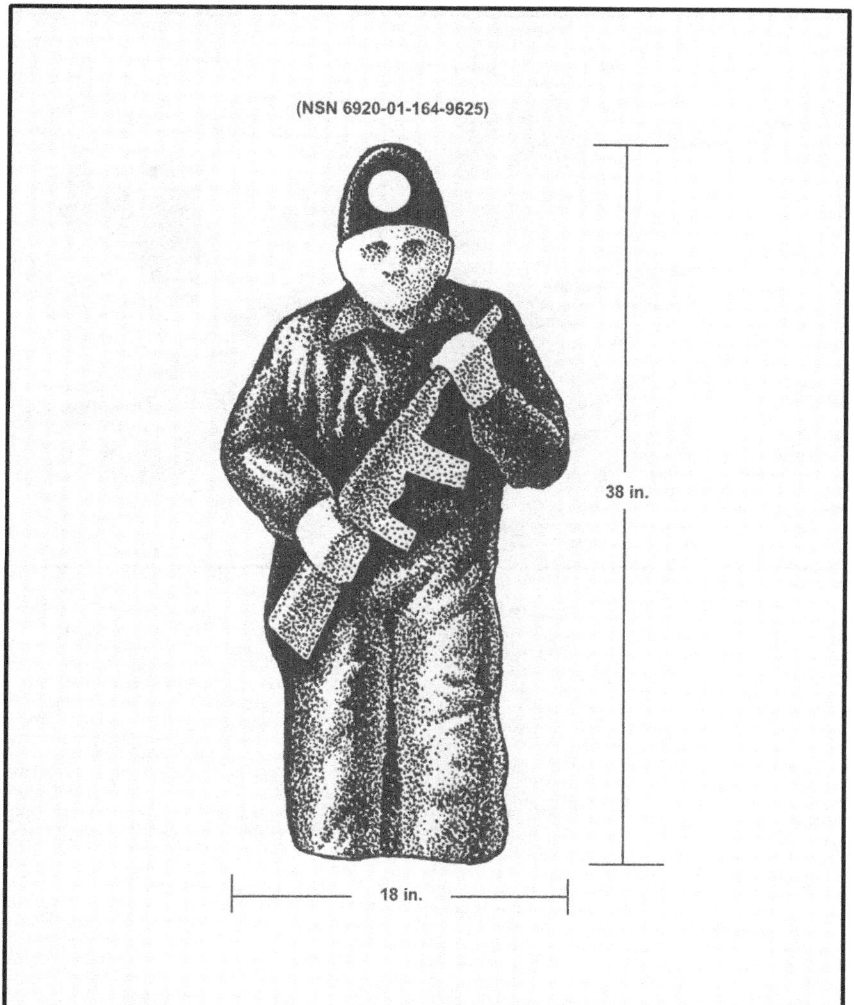


**Figure B-8. 3D personnel target**[1]

[1] Training requires 90% of the target be visible from the firing position.

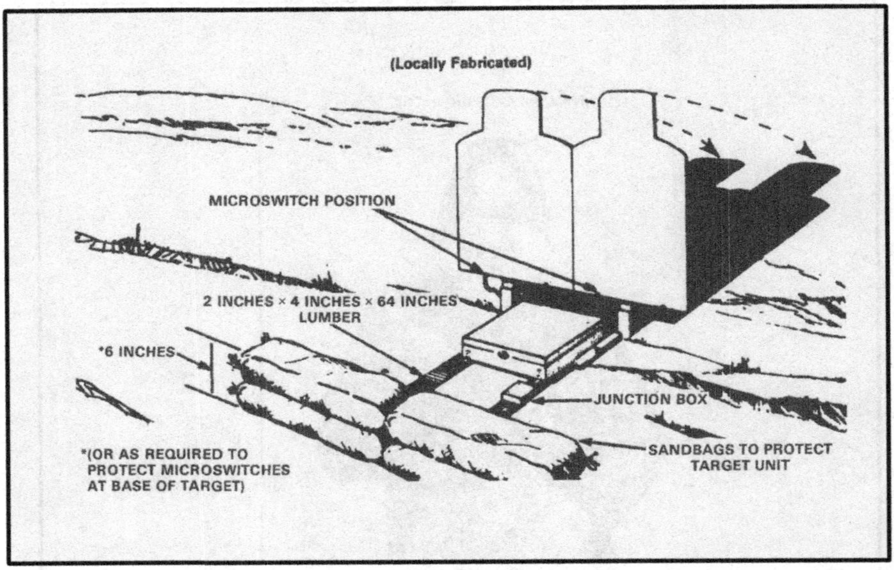


**Figure B-9. Small area target**

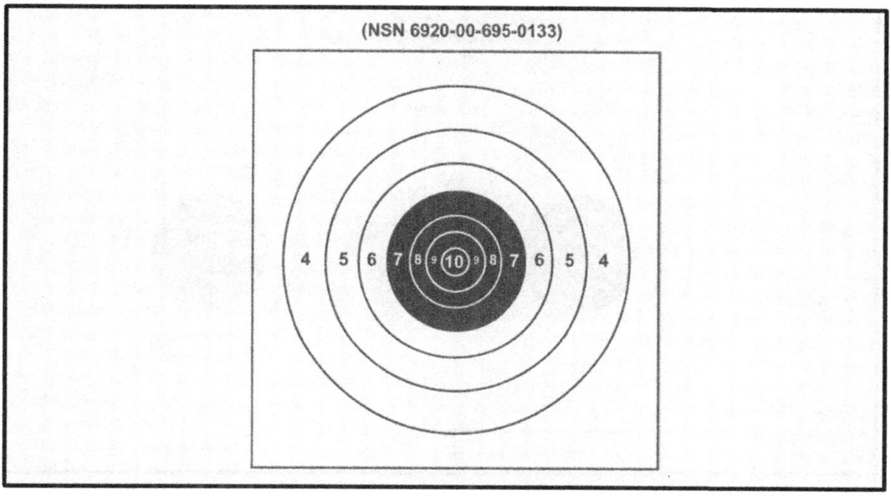


**Figure B-10. 50-foot slow-fire pistol bull's eye paper target**

**Figure B-11. 50-foot timed and rapid-fire pistol bull's eye paper target**

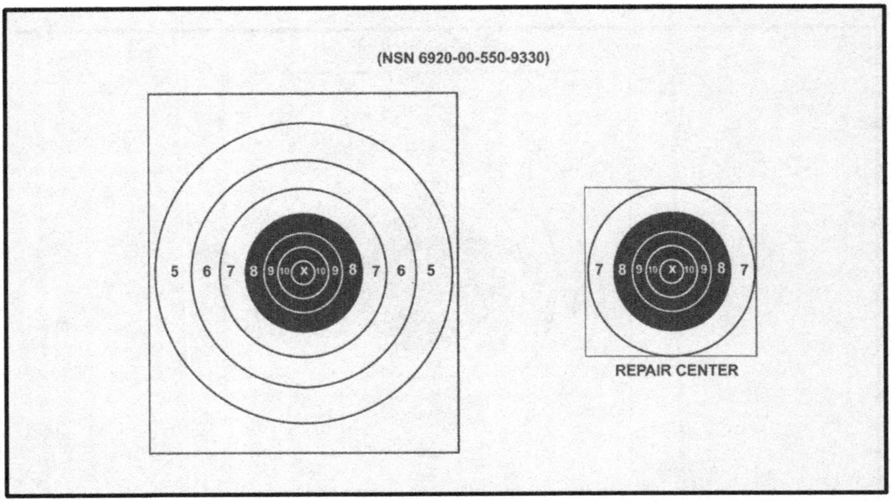


Figure B-12. 50-foot pistol target

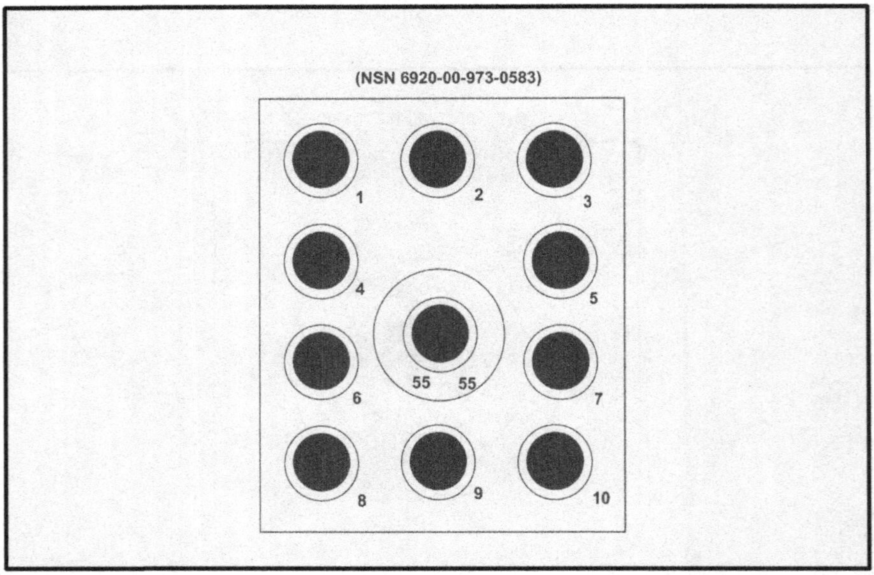


Figure B-13. 50-foot indoor rifle target

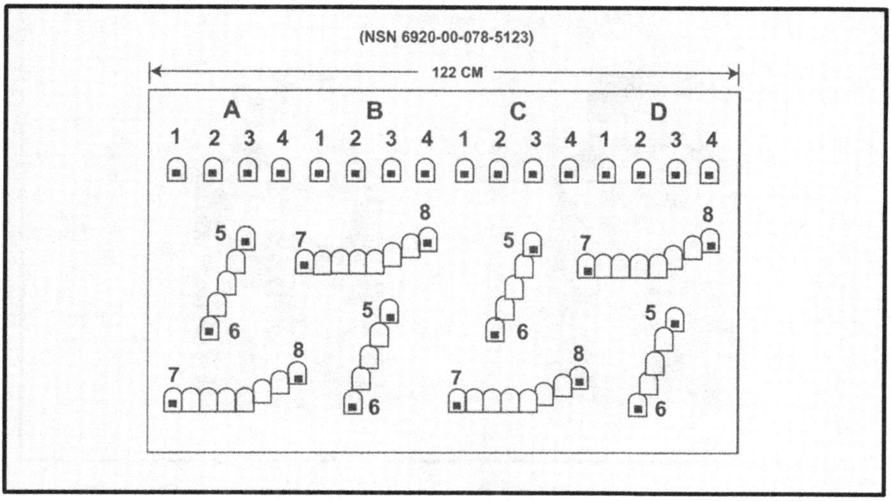


Figure B-14. Basic M60 machine gun target

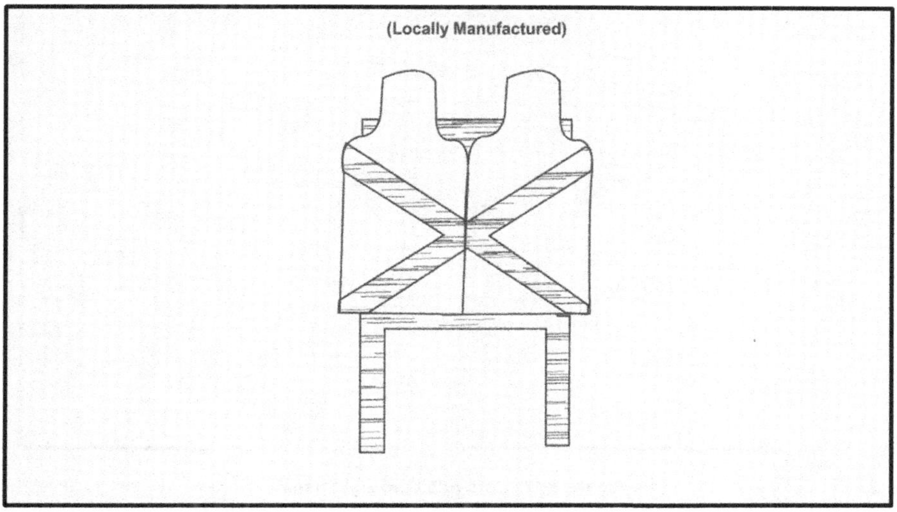


Figure B-15. Double E-type silhouette

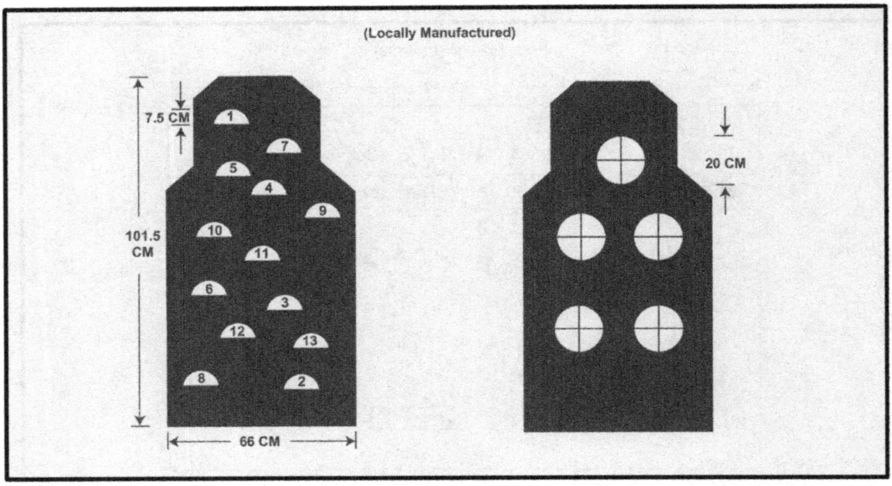


**Figure B-16. Target for zeroing exercises (Table I)**

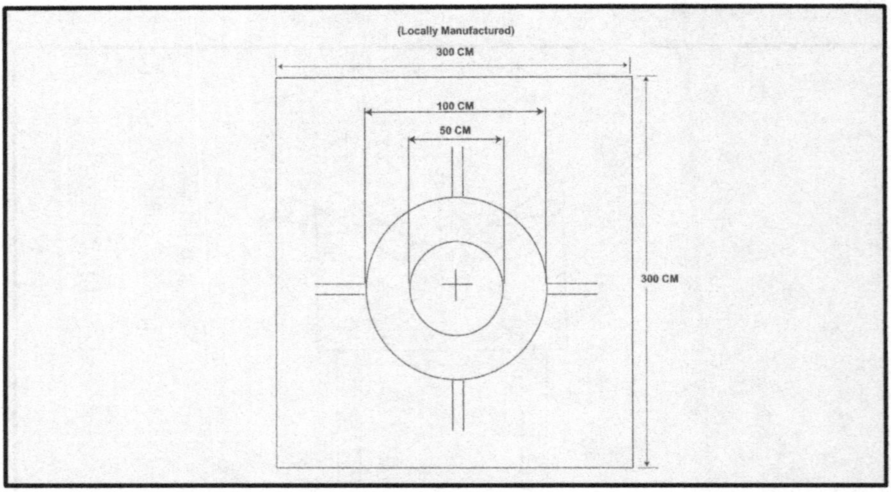


**Figure B-17. Zero and bore sight panel**

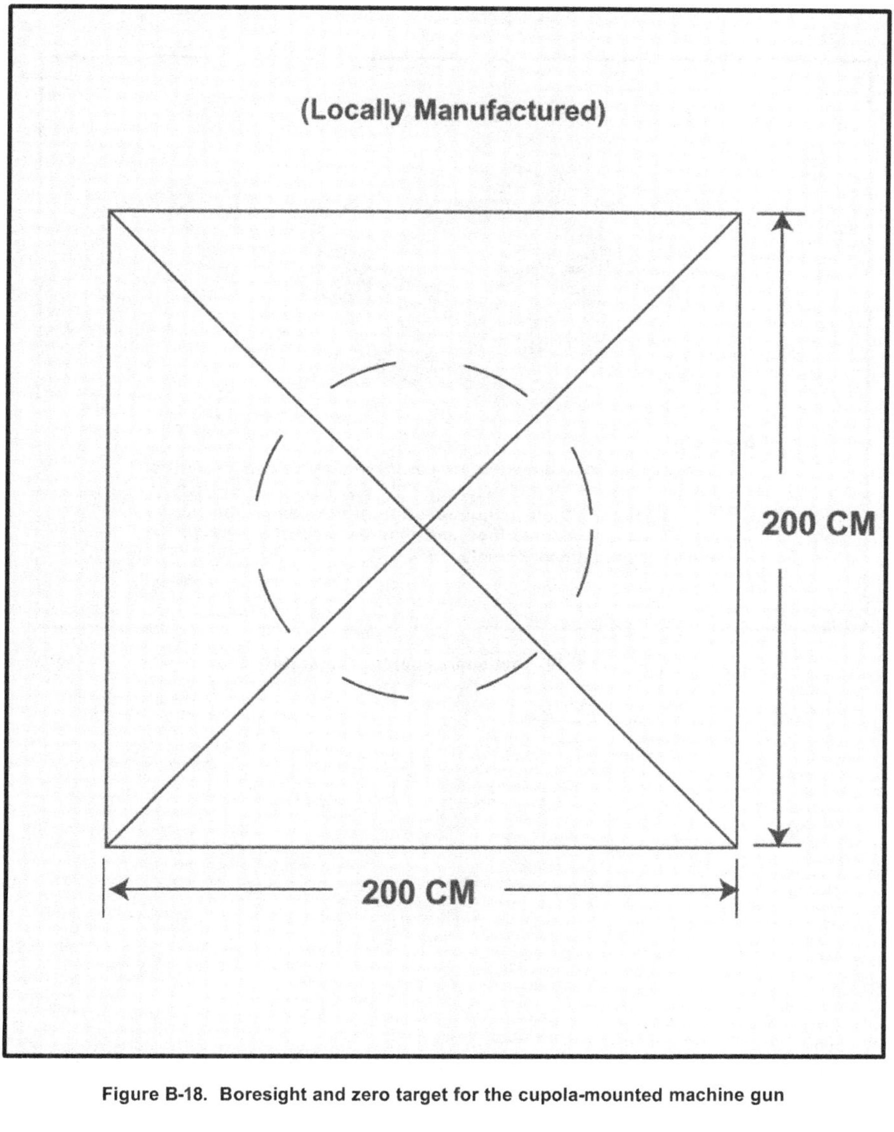


**Figure B-18. Boresight and zero target for the cupola-mounted machine gun**

(NSN 6920-01-2858)

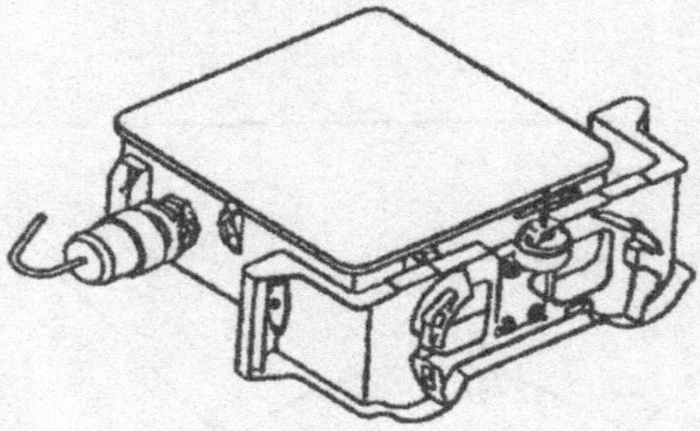

NOTE:
This mechanism mounts either two- or three-dimensional E and F silhouette targets. It raises a target in less than a second. It is automatically or manually controlled from the Range Control Station. It has laser interface and mounts on RETS moving target carrier or stationary emplacement.

**Figure B-19. Stationary Infantry Target (SIT)**

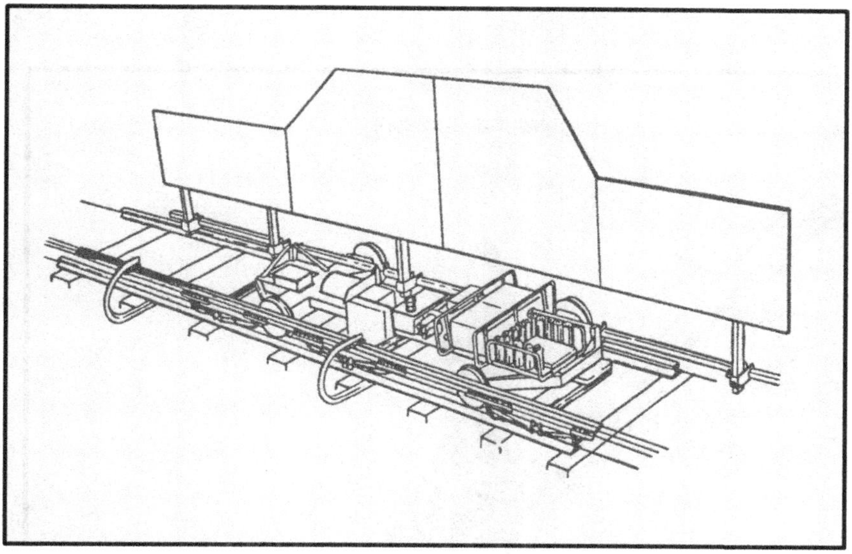

**Figure B-20. Moving Armor Target (MAT)**

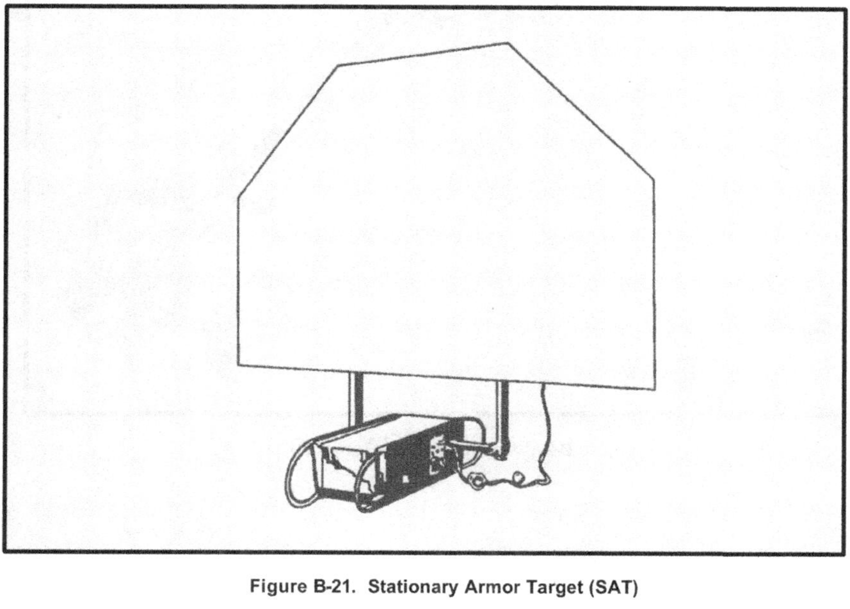

**Figure B-21. Stationary Armor Target (SAT)**

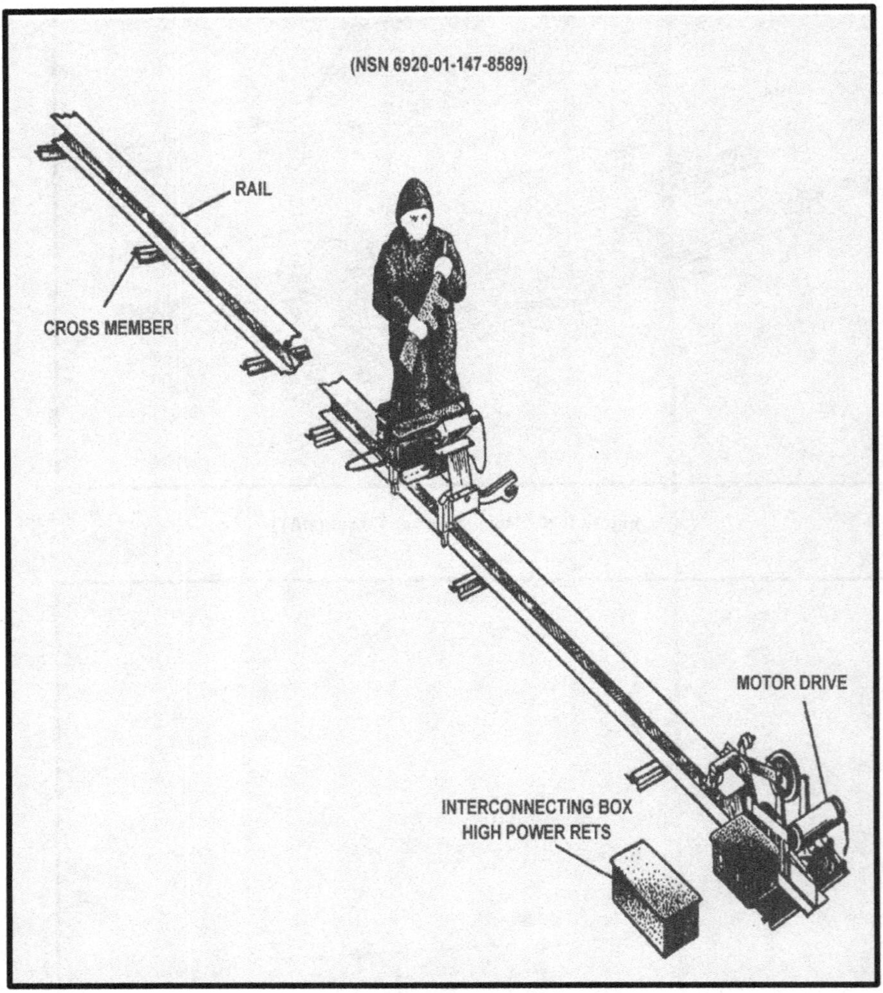


**Figure B-22. Moving Infantry Target (MIT)**

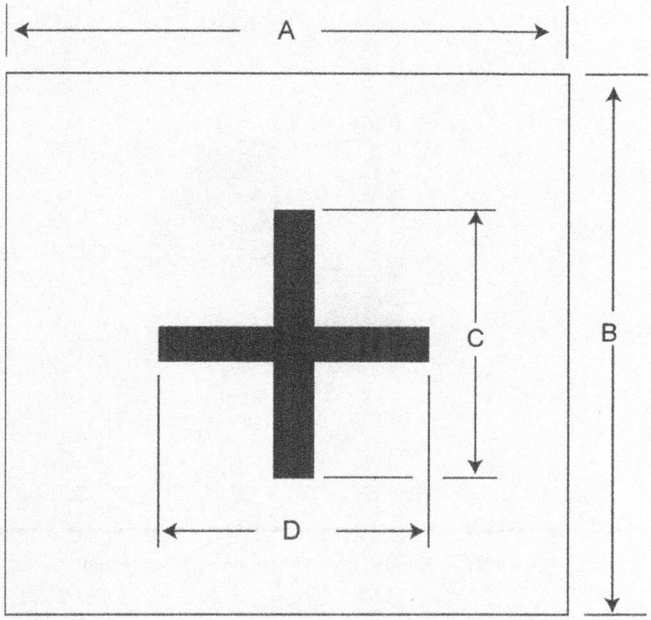


| | 1-2 METERS | INCHES |
|---|---|---|
| A | 3.0 | 118.0 |
| B | 3.0 | 118.0 |
| C | 0.6 | 23.6 |
| D | 0.6 | 23.6 |

**Figure B-23. Calibration panel (50 cal.)**

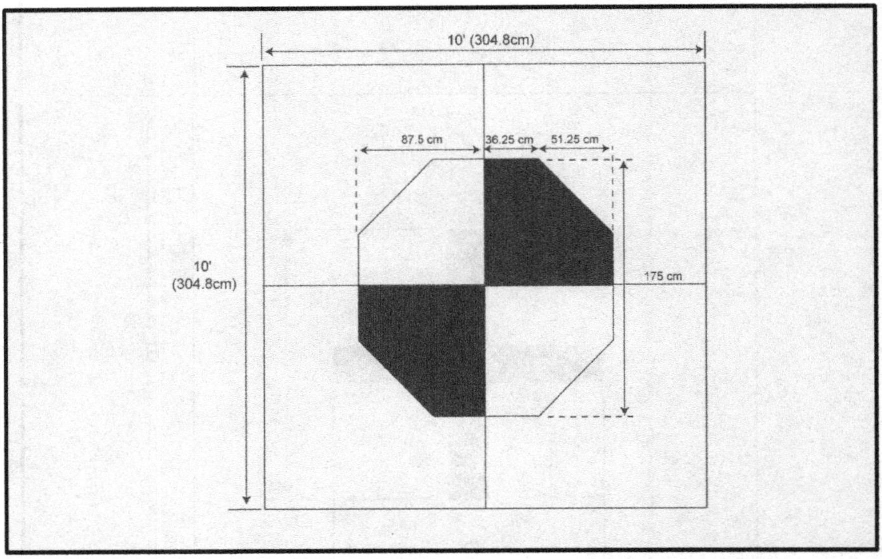


**Figure B-24. Screening test target no. 4 (120mm gun)**

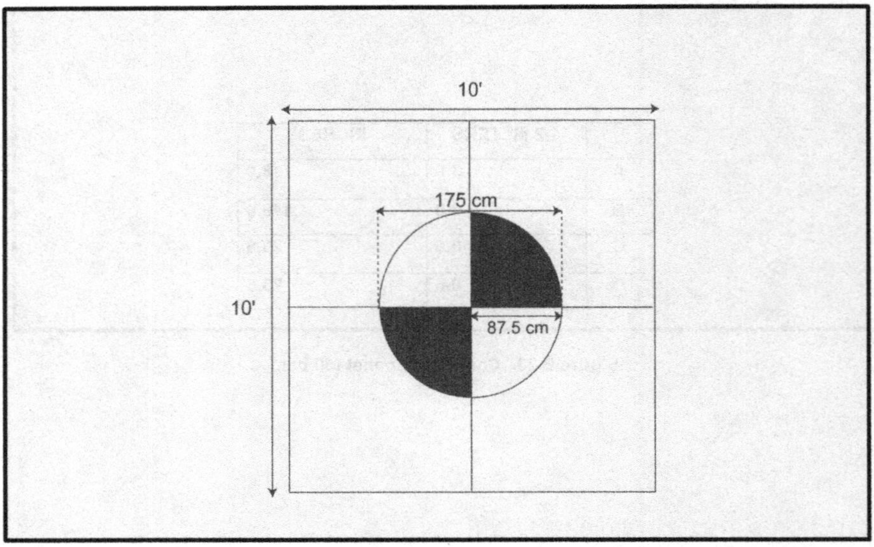


**Figure B-25. Screening test target no. 5 (120mm gun)**

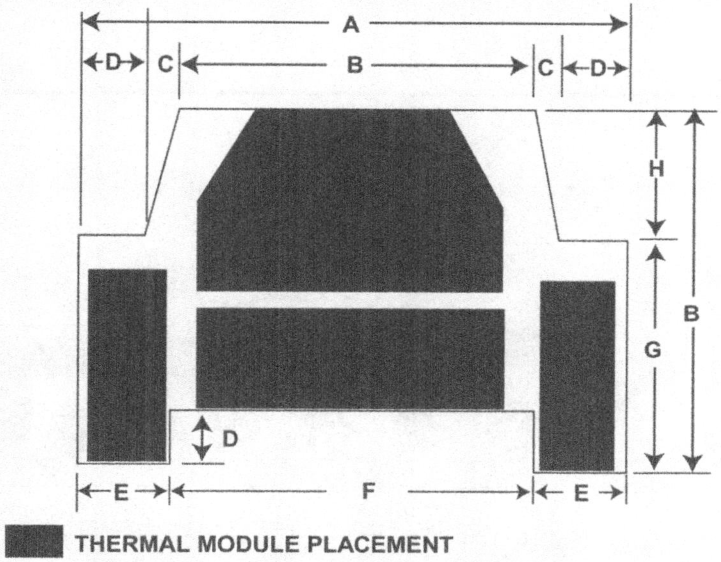


TARGET DIMENSIONS

| | FULL-SCALE | | 1:2 SCALE | 1:5 SCALE | 1:10 SCALE |
|---|---|---|---|---|---|
| | INCHES | CENTIMETERS | CENTIMETERS | CENTIMETERS | CENTIMETERS |
| A | 132.0 | 335.0 | 167.5 | 67.0 | 33.5 |
| B | 87.0 | 220.0 | 110.0 | 44.0 | 22.0 |
| C | 6.5 | 17.5 | 8.75 | 3.5 | 1.75 |
| D | 16.0 | 40.0 | 20.0 | 8.0 | 4.0 |
| E | 20.0 | 50.0 | 25.0 | 10.0 | 5.0 |
| F | 92.0 | 235.0 | 117.5 | 47.0 | 23.5 |
| G | 55.0 | 140.0 | 70.0 | 28.0 | 14.0 |
| H | 32.0 | 80.0 | 40.0 | 16.0 | 8.0 |

**Figure B-26. T-72 front target**

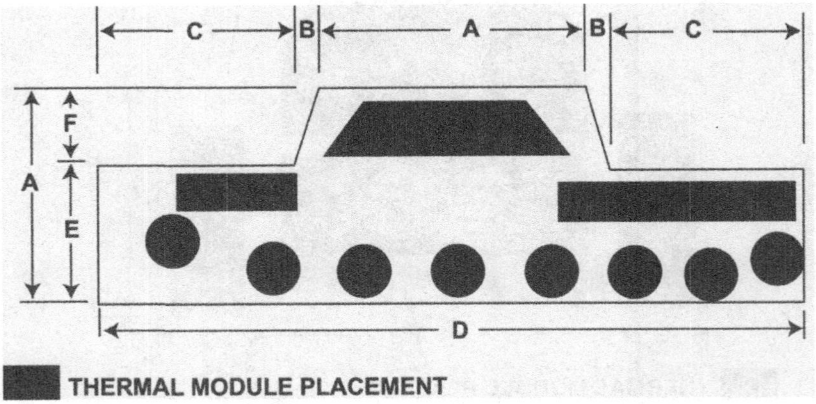


TARGET DIMENSIONS

| | FULL-SCALE | | 1:2 SCALE | 1:5 SCALE | 1:10 SCALE |
|---|---|---|---|---|---|
| | INCHES | CENTIMETERS | CENTIMETERS | CENTIMETERS | CENTIMETERS |
| A | 90.5 | 230.0 | 115.0 | 46.0 | 23.0 |
| B | 16.0 | 40.0 | 20.0 | 8.0 | 4.0 |
| C | 79.0 | 200.0 | 100.0 | 40.0 | 20.0 |
| D | 276.0 | 700.0 | 350.0 | 140.0 | 70.0 |
| E | 55.0 | 140.0 | 70.0 | 28.0 | 14.0 |
| F | 35.0 | 90.0 | 45.0 | 18.0 | 9.0 |

**Figure B-27. T-72 flank target**

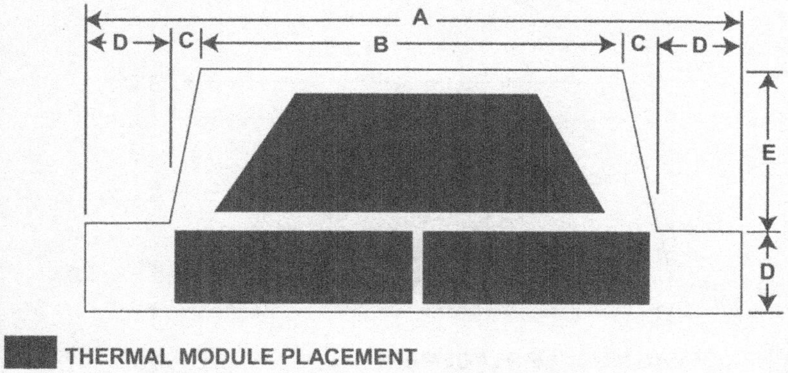


TARGET DIMENSIONS

| | FULL-SCALE | | 1:2 SCALE | 1:5 SCALE | 1:10 SCALE |
|---|---|---|---|---|---|
| | INCHES | CENTIMETERS | CENTIMETERS | CENTIMETERS | CENTIMETERS |
| A | 132.0 | 335.0 | 167.5 | 67.0 | 33.5 |
| B | 87.0 | 220.0 | 110.0 | 44.0 | 22.0 |
| C | 6.5 | 17.5 | 8.75 | 3.5 | 1.7 |
| D | 16.0 | 40.0 | 20.0 | 8.0 | 4.0 |
| E | 35.0 | 90.0 | 45.0 | 18.0 | 9.0 |

**Figure B-28. T-72 partial defilade target**

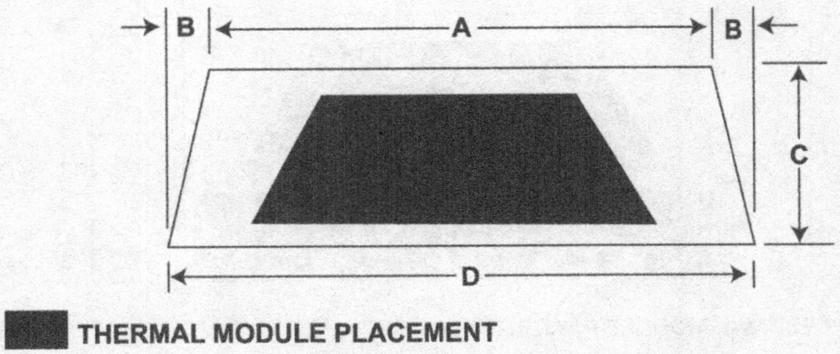


TARGET DIMENSIONS

| | FULL-SCALE INCHES | FULL-SCALE CENTIMETERS | 1:2 SCALE CENTIMETERS | 1:5 SCALE CENTIMETERS | 1:10 SCALE CENTIMETERS |
|---|---|---|---|---|---|
| A | 87.0 | 220.0 | 110.0 | 44.0 | 22.0 |
| B | 5.5 | 15.0 | 7.5 | 3.0 | 1.5 |
| C | 35.0 | 90.0 | 45.0 | 18.0 | 9.0 |
| D | 98.0 | 250.0 | 125.0 | 50.0 | 25.0 |

**Figure B-29. T-72 Turret Target**

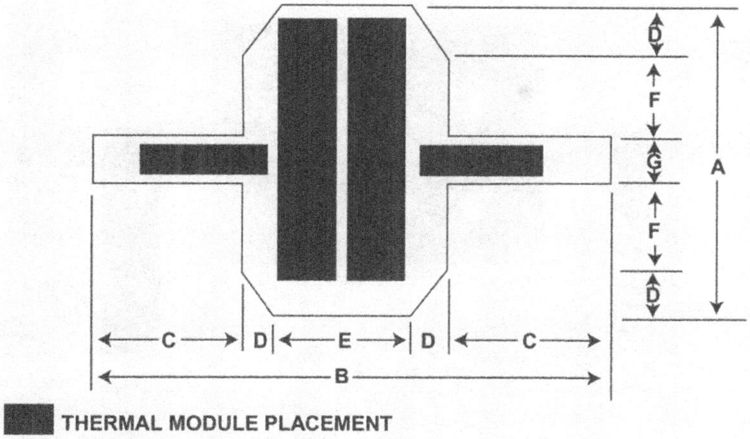


TARGET DIMENSIONS

| | FULL-SCALE | | 1:2 SCALE | 1:5 SCALE | 1:10 SCALE |
|---|---|---|---|---|---|
| | INCHES | CENTIMETERS | CENTIMETERS | CENTIMETERS | CENTIMETERS |
| A | 120.0 | 305.0 | 152.5 | 61.0 | 30.5 |
| B | 192.0 | 488.0 | 244.0 | 97.6 | 48.8 |
| C | 60.0 | 152.0 | 76.0 | 30.4 | 15.2 |
| D | 12.0 | 31.0 | 15.5 | 6.2 | 3.1 |
| E | 48.0 | 122.0 | 61.0 | 24.4 | 12.2 |
| F | 36.0 | 91.0 | 45.5 | 18.2 | 9.1 |
| G | 24.0 | 61.0 | 30.5 | 12.2 | 6.1 |

**Figure B-30. HIND front target**

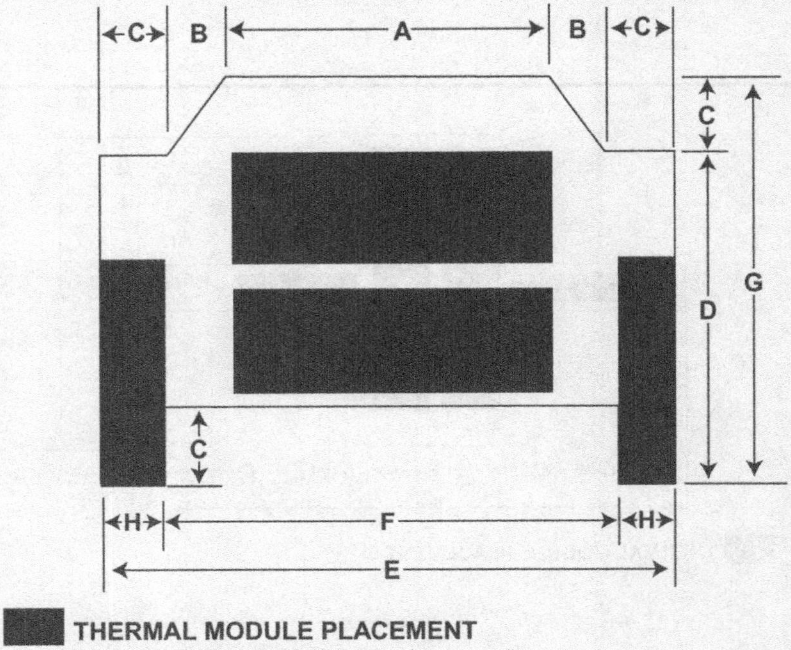


TARGET DIMENSIONS

| | FULL-SCALE | | 1:2 SCALE | 1:5 SCALE | 1:10 SCALE |
|---|---|---|---|---|---|
| | INCHES | CENTIMETERS | CENTIMETERS | CENTIMETERS | CENTIMETERS |
| A | 70.0 | 179.0 | 89.5 | 35.8 | 17.9 |
| B | 10.0 | 25.5 | 12.75 | 5.1 | 2.6 |
| C | 16.0 | 40.0 | 20.0 | 8.0 | 4.0 |
| D | 71.0 | 180.0 | 90.0 | 36.0 | 18.0 |
| E | 122.0 | 310.0 | 155.0 | 62.0 | 31.0 |
| F | 98.0 | 250.0 | 125.0 | 50.0 | 25.0 |
| G | 87.0 | 220.0 | 110.0 | 44.0 | 22.0 |
| H | 12.0 | 30.0 | 15.0 | 6.0 | 3.0 |

**Figure B-31. BMP 1981 front target**

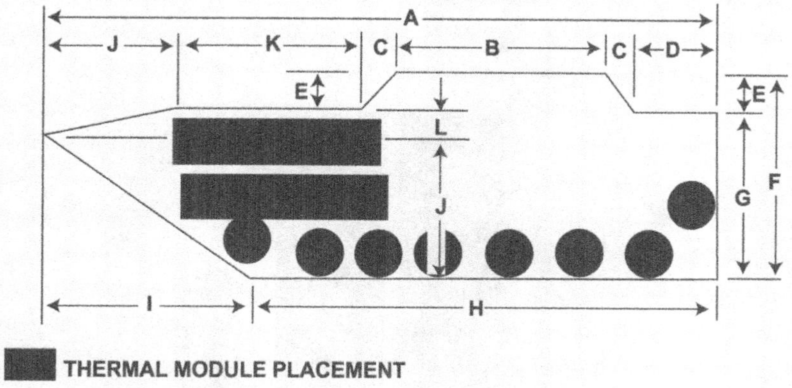


TARGET DIMENSIONS

| | FULL-SCALE | | 1:2 SCALE | 1:5 SCALE | 1:10 SCALE |
|---|---|---|---|---|---|
| | INCHES | CENTIMETERS | CENTIMETERS | CENTIMETERS | CENTIMETERS |
| A | 268.0 | 680.0 | 340.0 | 136.0 | 68.0 |
| B | 67.0 | 170.0 | 85.0 | 34.0 | 17.0 |
| C | 12.0 | 30.0 | 15.0 | 6.0 | 3.0 |
| D | 39.0 | 100.0 | 50.0 | 20.0 | 10.0 |
| E | 18.0 | 45.0 | 22.5 | 9.0 | 4.5 |
| F | 87.0 | 220.0 | 110.0 | 44.0 | 22.0 |
| G | 69.0 | 175.0 | 87.5 | 35.0 | 17.5 |
| H | 173.0 | 440.0 | 220.0 | 88.0 | 44.0 |
| I | 95.0 | 240.0 | 120.0 | 48.0 | 24.0 |
| J | 61.0 | 155.0 | 77.5 | 31.0 | 15.5 |
| K | 77.0 | 195.0 | 97.5 | 39.0 | 19.5 |
| L | 8.0 | 20.0 | 10.0 | 4.0 | 2.0 |

**Figure B-32. BMP 1981 flank target**

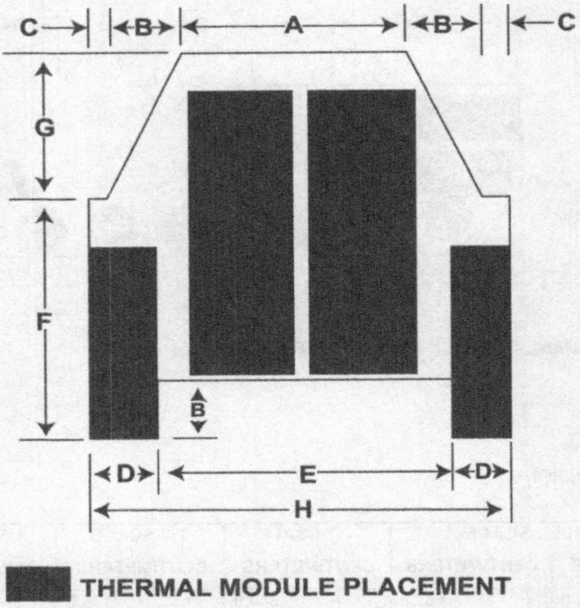


TARGET DIMENSIONS

| | FULL-SCALE | | 1:2 SCALE | 1:5 SCALE | 1:10 SCALE |
|---|---|---|---|---|---|
| | INCHES | CENTIMETERS | CENTIMETERS | CENTIMETERS | CENTIMETERS |
| A | 47.0 | 120.0 | 60.0 | 24.0 | 12.0 |
| B | 16.0 | 40.0 | 20.0 | 8.0 | 4.0 |
| C | 4.0 | 10.0 | 5.0 | 2.0 | 1.0 |
| D | 12.0 | 30.0 | 15.0 | 6.0 | 3.0 |
| E | 63.0 | 160.0 | 80.0 | 32.0 | 16.0 |
| F | 59.0 | 150.0 | 75.0 | 30.0 | 15.0 |
| G | 35.0 | 90.0 | 45.0 | 18.0 | 9.0 |
| H | 87.0 | 220.0 | 110.0 | 44.0 | 22.0 |

**Figure B-33. BRDM front target**

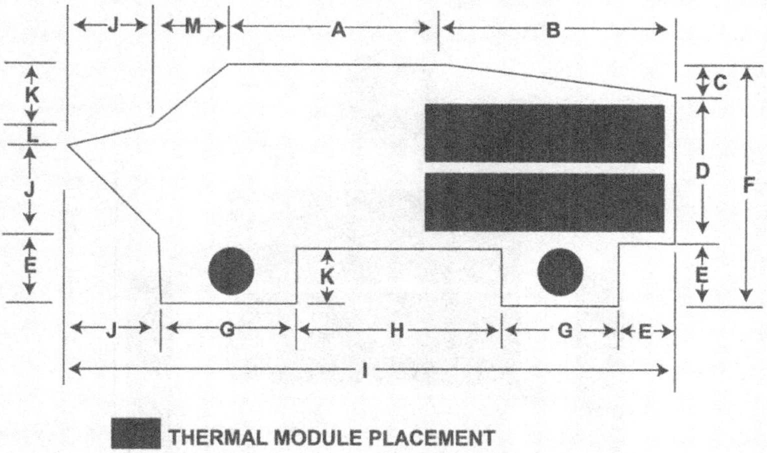


TARGET DIMENSIONS

| | FULL-SCALE | | 1:2 SCALE | 1:5 SCALE | 1:10 SCALE |
|---|---|---|---|---|---|
| | INCHES | CENTIMETERS | CENTIMETERS | CENTIMETERS | CENTIMETERS |
| A | 87.0 | 221.0 | 110.5 | 44.2 | 22.1 |
| B | 81.0 | 206.0 | 103.0 | 41.2 | 20.6 |
| C | 11.0 | 28.0 | 14.0 | 5.6 | 2.8 |
| D | 48.0 | 122.0 | 61.0 | 24.4 | 12.2 |
| E | 24.0 | 61.0 | 30.5 | 12.2 | 6.1 |
| F | 83.0 | 211.0 | 105.5 | 42.2 | 21.1 |
| G | 47.0 | 120.0 | 60.0 | 24.0 | 12.0 |
| H | 75.0 | 190.0 | 95.0 | 38.0 | 19.0 |
| I | 224.0 | 570.0 | 285.0 | 114.0 | 57.0 |
| J | 31.0 | 79.0 | 39.5 | 15.8 | 7.9 |
| K | 20.0 | 51.0 | 25.5 | 10.2 | 5.1 |
| L | 8.0 | 20.0 | 10.0 | 4.0 | 2.0 |
| M | 25.0 | 64.0 | 32.0 | 12.8 | 6.4 |

**Figure B-34. BRDM flank target**

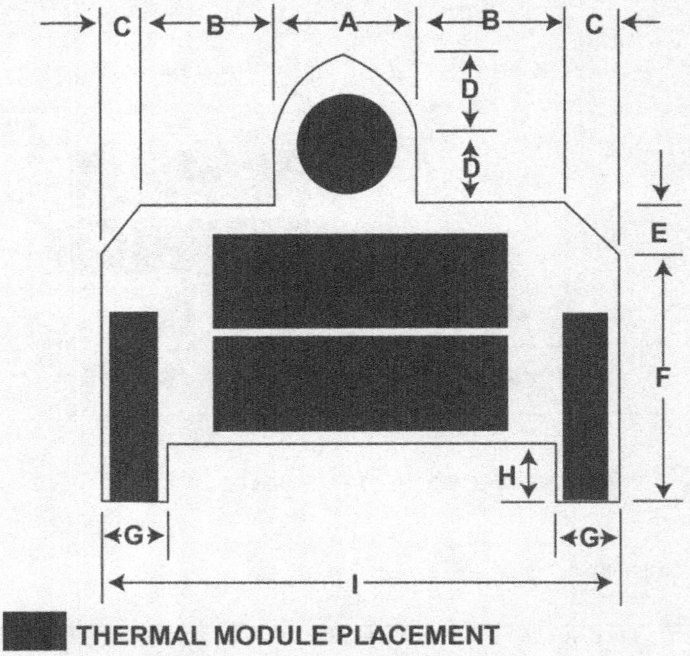


TARGET DIMENSIONS

| | FULL-SCALE | | 1:2 SCALE | 1:5 SCALE | 1:10 SCALE |
|---|---|---|---|---|---|
| | INCHES | CENTIMETERS | CENTIMETERS | CENTIMETERS | CENTIMETERS |
| A | 35.0 | 90.0 | 45.0 | 18.0 | 9.0 |
| B | 33.5 | 85.0 | 42.5 | 17.0 | 8.5 |
| C | 8.0 | 20.0 | 10.0 | 4.0 | 2.0 |
| D | 20.0 | 50.0 | 25.0 | 10.0 | 5.0 |
| E | 12.0 | 30.0 | 15.0 | 6.0 | 3.0 |
| F | 59.0 | 150.0 | 75.0 | 30.0 | 15.0 |
| G | 14.0 | 35.0 | 17.5 | 7.0 | 3.5 |
| H | 16.0 | 40.0 | 20.0 | 8.0 | 4.0 |
| I | 118.0 | 300.0 | 150.0 | 60.0 | 30.0 |

**Figure B-35. ZSU-23/4 front target**

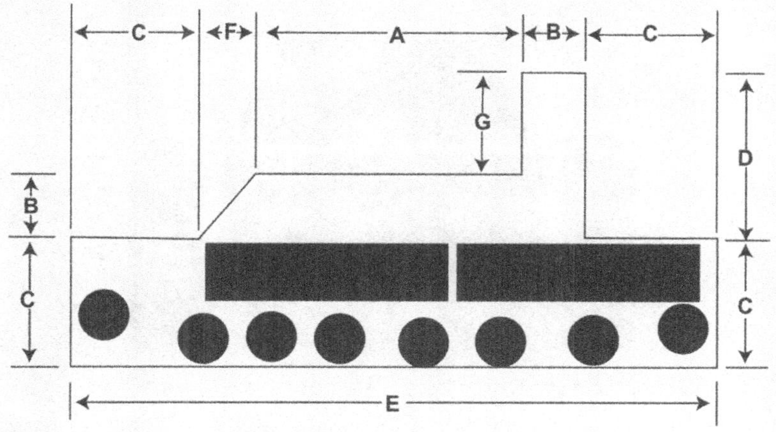


TARGET DIMENSIONS

| | FULL-SCALE | | 1:2 SCALE | 1:5 SCALE | 1:10 SCALE |
|---|---|---|---|---|---|
| | INCHES | CENTIMETERS | CENTIMETERS | CENTIMETERS | CENTIMETERS |
| A | 98.0 | 250.0 | 125.0 | 50.0 | 25.0 |
| B | 24.0 | 60.0 | 30.0 | 12.0 | 60.0 |
| C | 47.0 | 120.0 | 60.0 | 24.0 | 120.0 |
| D | 63.0 | 160.0 | 80.0 | 32.0 | 16.0 |
| E | 236.0 | 600.0 | 300.0 | 120.0 | 60.0 |
| F | 20.0 | 50.0 | 25.0 | 10.0 | 5.0 |
| G | 39.0 | 100.0 | 50.0 | 20.0 | 10.0 |

**Figure B-36. ZSU-23/4 flank target**

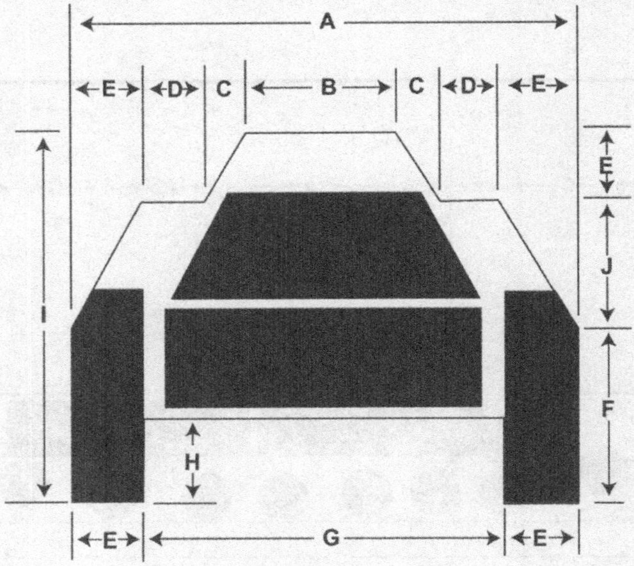


TARGET DIMENSIONS

| | FULL-SCALE | | 1:2 SCALE | 1:5 SCALE | 1:10 SCALE |
|---|---|---|---|---|---|
| | INCHES | CENTIMETERS | CENTIMETERS | CENTIMETERS | CENTIMETERS |
| A | 110.0 | 280.0 | 140.0 | 56.0 | 28.0 |
| B | 30.0 | 800.0 | 40.0 | 16.0 | 8.0 |
| C | 10.0 | 25.0 | 12.5 | 5.0 | 2.5 |
| D | 14.0 | 35.0 | 17.5 | 7.0 | 3.5 |
| E | 16.0 | 40.0 | 20.0 | 8.0 | 4.0 |
| F | 39.0 | 100.0 | 50.0 | 20.0 | 10.0 |
| G | 78.0 | 200.0 | 100.0 | 40.0 | 20.0 |
| H | 20.0 | 50.0 | 25.0 | 10.0 | 5.0 |
| I | 86.0 | 220.0 | 110.0 | 44.0 | 22.0 |
| J | 31.0 | 80.0 | 40.0 | 16.0 | 8.0 |

**Figure B-37. BTR-series front target**

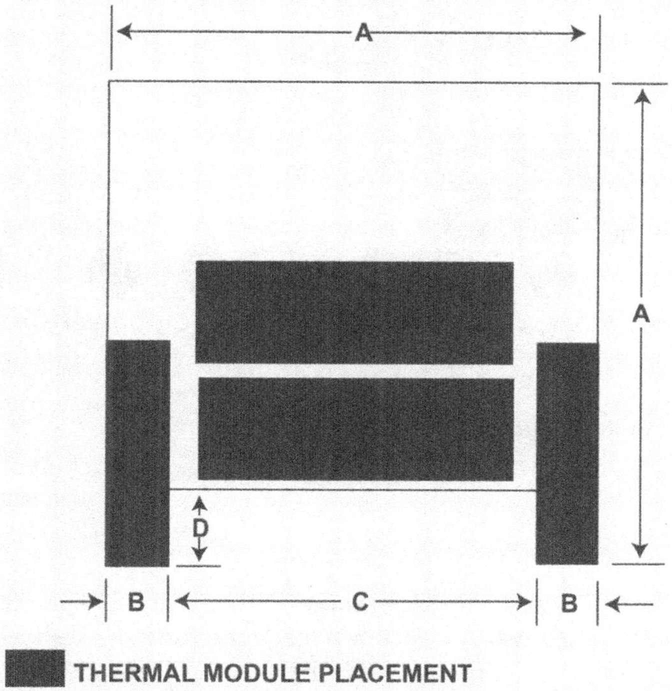


TARGET DIMENSIONS

| | FULL-SCALE | | 1:2 SCALE | 1:5 SCALE | 1:10 SCALE |
|---|---|---|---|---|---|
| | INCHES | CENTIMETERS | CENTIMETERS | CENTIMETERS | CENTIMETERS |
| A | 102.0 | 260.0 | 130.0 | 52.0 | 26.0 |
| B | 12.0 | 30.0 | 15.0 | 6.0 | 3.0 |
| C | 78.0 | 200.0 | 100.0 | 40.0 | 20.0 |
| D | 16.0 | 40.0 | 20.0 | 8.0 | 4.0 |

**Figure B-38. Truck U-375 front target**

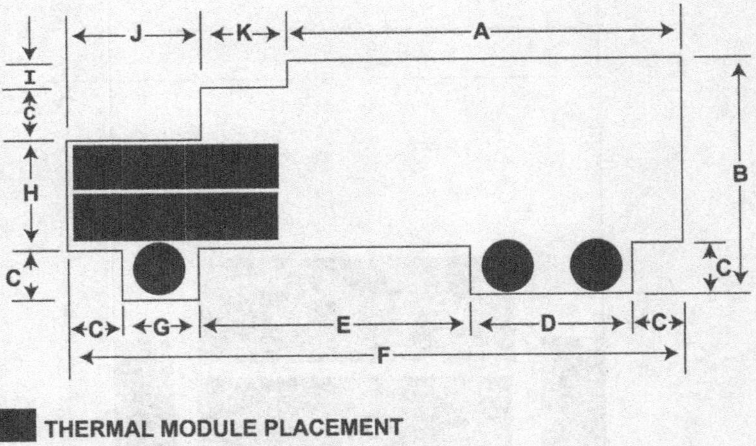


TARGET DIMENSIONS

| | FULL-SCALE | | 1:2 SCALE | 1:5 SCALE | 1:10 SCALE |
|---|---|---|---|---|---|
| | INCHES | CENTIMETERS | CENTIMETERS | CENTIMETERS | CENTIMETERS |
| A | 177.0 | 450.0 | 225.0 | 90.0 | 45.0 |
| B | 102.0 | 260.0 | 130.0 | 52.0 | 26.0 |
| C | 24.0 | 60.0 | 30.0 | 12.0 | 6.0 |
| D | 71.0 | 180.0 | 90.0 | 36.0 | 18.0 |
| E | 126.0 | 320.0 | 160.0 | 64.0 | 32.0 |
| F | 276.0 | 700.0 | 350.0 | 140.0 | 70.0 |
| G | 31.0 | 80.0 | 40.0 | 16.0 | 8.0 |
| H | 47.0 | 120.0 | 60.0 | 24.0 | 12.0 |
| I | 7.0 | 20.0 | 100.0 | 4.0 | 2.0 |
| J | 59.0 | 150.0 | 750.0 | 30.0 | 15.0 |
| K | 40.0 | 100.0 | 50.0 | 20.0 | 10.0 |

**Figure B-39. Truck U-375 flank target**

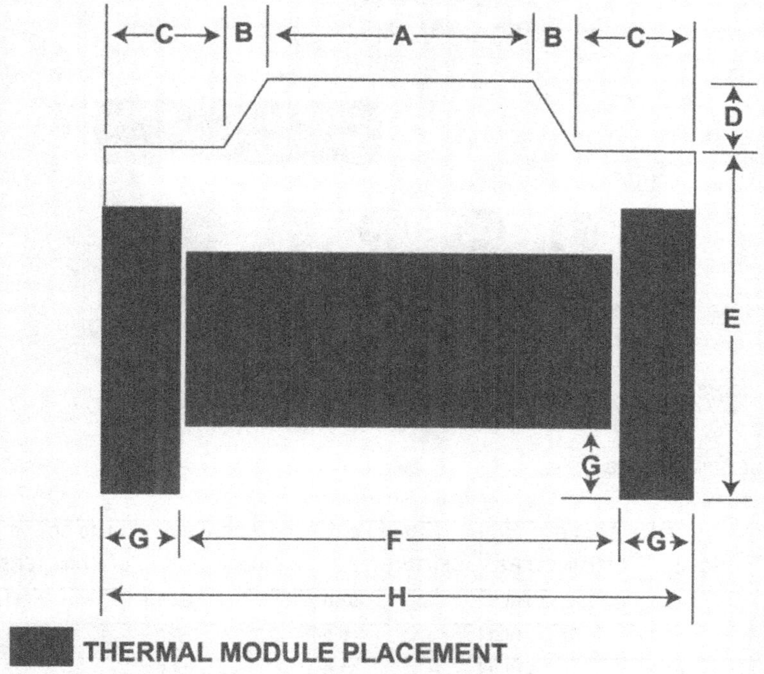


TARGET DIMENSIONS

| | FULL-SCALE | | 1:2 SCALE | 1:5 SCALE | 1:10 SCALE |
|---|---|---|---|---|---|
| | INCHES | CENTIMETERS | CENTIMETERS | CENTIMETERS | CENTIMETERS |
| A | 43.0 | 110.0 | 55.0 | 22.0 | 11.0 |
| B | 80.0 | 20.0 | 10.0 | 4.0 | 2.0 |
| C | 20.0 | 50.0 | 25.0 | 10.0 | 5.0 |
| D | 12.0 | 30.0 | 15.0 | 6.0 | 3.0 |
| E | 55.0 | 140.0 | 70.0 | 28.0 | 14.0 |
| F | 67.0 | 168.0 | 84.0 | 33.6 | 16.8 |
| G | 16.0 | 40.0 | 20.0 | 8.0 | 4.0 |
| H | 99.0 | 250.0 | 125.0 | 50.0 | 25.0 |

**Figure B-40. BMD front target**

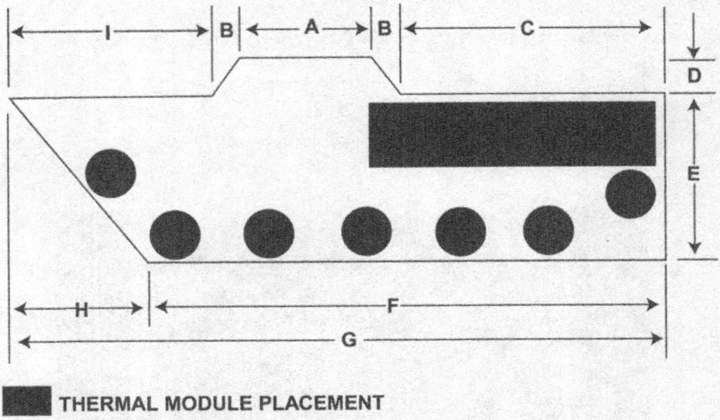


TARGET DIMENSIONS

| | FULL-SCALE | | 1:2 SCALE | 1:5 SCALE | 1:10 SCALE |
|---|---|---|---|---|---|
| | INCHES | CENTIMETERS | CENTIMETERS | CENTIMETERS | CENTIMETERS |
| A | 45.0 | 114.0 | 57.0 | 22.8 | 11.4 |
| B | 7.0 | 18.0 | 9.0 | 3.6 | 1.8 |
| C | 87.0 | 220.0 | 11.0 | 44.0 | 22.0 |
| D | 12.0 | 30.0 | 15.0 | 6.0 | 3.0 |
| E | 55.0 | 140.0 | 70.0 | 28.0 | 14.0 |
| F | 173.0 | 438.0 | 219.0 | 87.6 | 43.8 |
| G | 213.0 | 540.0 | 270.0 | 108.0 | 54.0 |
| H | 400.0 | 102.0 | 51.0 | 20.4 | 10.2 |
| I | 67.0 | 1700.0 | 85.0 | 34.0 | 17.0 |

**Figure B-41. BMD flank target**

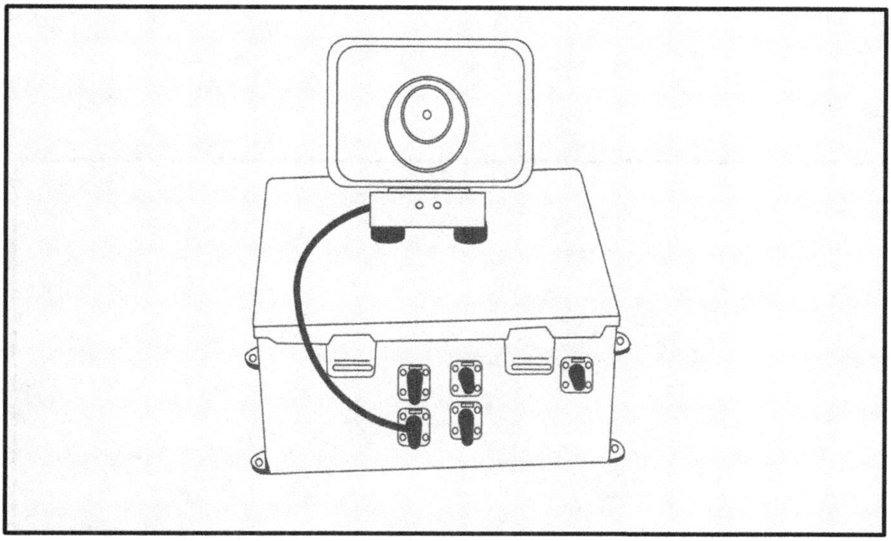

Figure B-42. Sound effects simulator (SES)

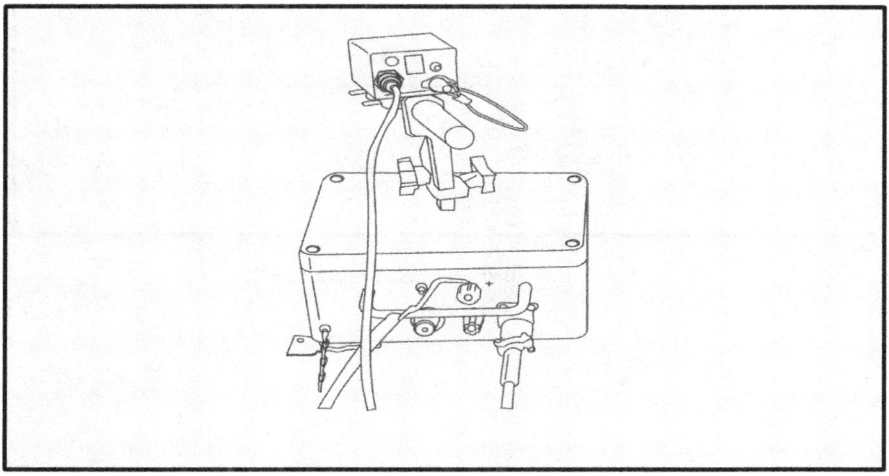

Figure B-43. MILES shootback device (MSD)

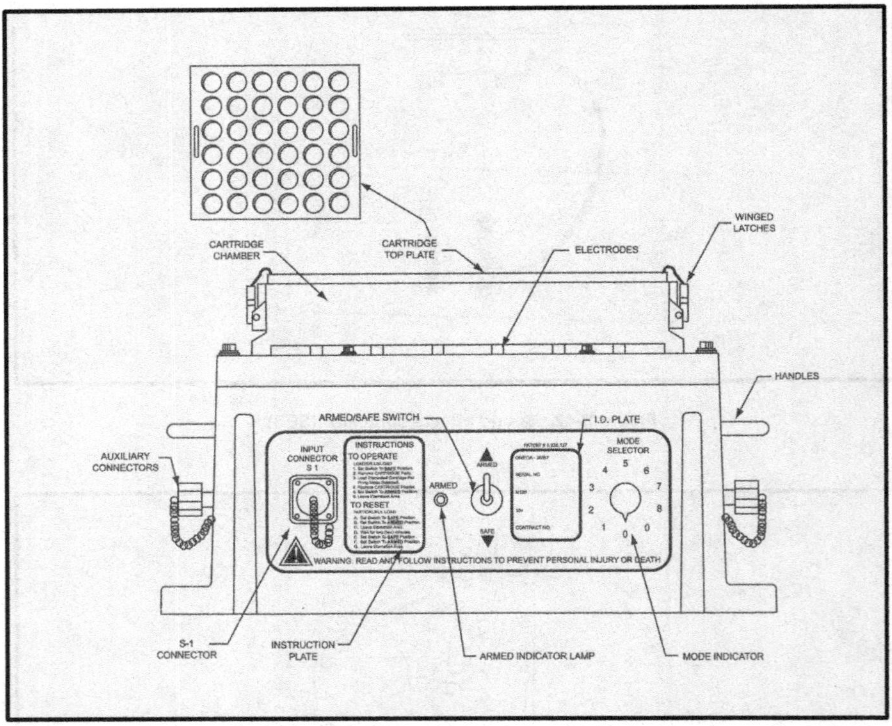


**Figure B-44. Battle effects simulator (BES)**

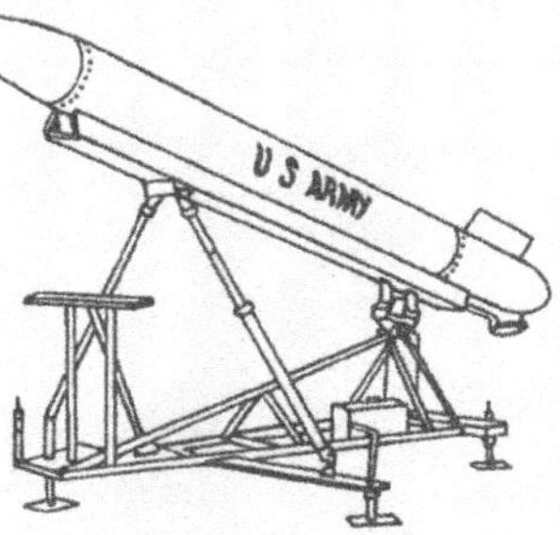


**NOTE:** BATS is a low-cost, one-time flight (expendable) target that is operated from the ground. It provides the firing unit with an on-call target and simulates the trajectory of a high-performance threat aircraft. BATS is used in support of STINGER training. Its characteristics:

Speed – 550 knots

Altitude – 8,000 feet

Endurance – 47 seconds

Figure B-45. Ballistic aerial target system (BATS)

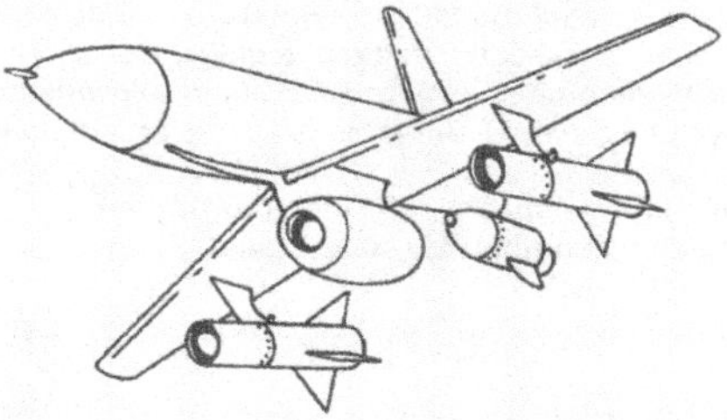

**NOTE:** The Streaker is a drone target that provides the altitude, speed, and maneuver capability required for supporting the training of Hawk units. It is a low-cost, air transportable system which can be used in both domestic and overseas training areas. The target has the following characteristics:

Speed – 250 to 500 knots

Altitude – 100 to 12,200 meters

Figure B-46. MQM-107 variable-speed training target Streaker

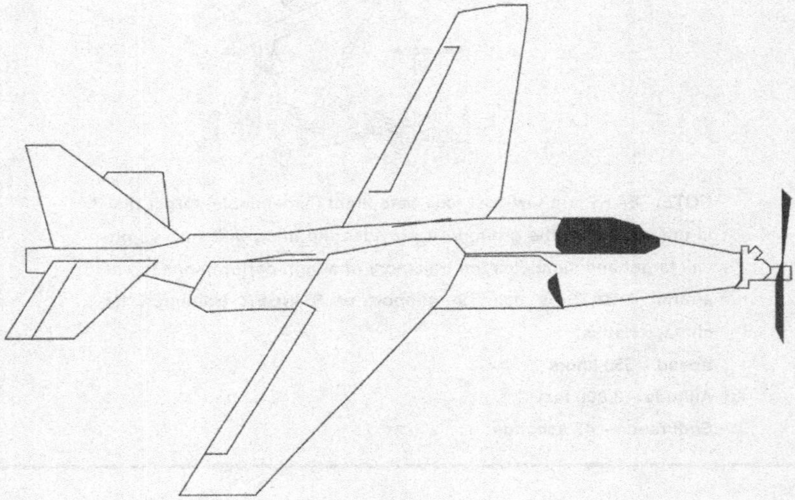

**NOTE: The subscale Su-25 Frogfoot is a highly maneuverable aircraft capable of executing attack profiles. It is recoverable in unimproved landing sites and reusable if not expended by live fire. It has a 20-pound payload and is controllable at a distance of three kilometers.**

**Speed: 45 mph (minimum), 100 mph (maximum)**

**Endurance: 30 minutes, maximum speed**

**Figure B-47. Su-25 Frogfoot aerial target**

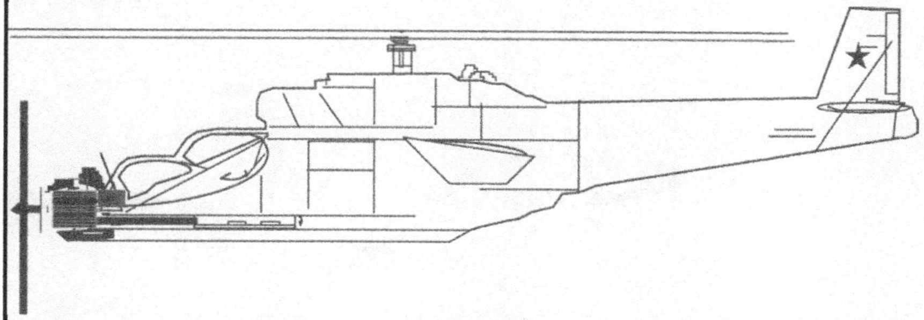

NOTE: The subscale Mi-24 HIND is a highly maneuverable aircraft capable of executing attack profiles. It is recoverable in unimproved landing sites and reusable if not expended by live fire. It has a 20-pound payload and is controllable at a distance of three kilometers.

Speed: 45 mph (minimum), 70 mph (maximum)
Endurance: 30 minutes, maximum speed

Figure B-48. Mi-24 HIND aerial target

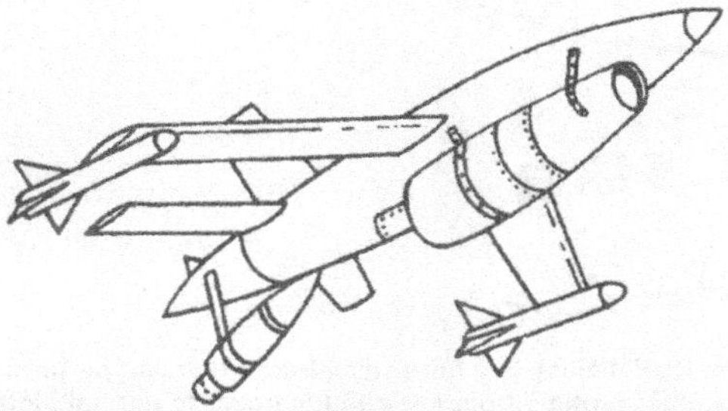

NOTE: The Firebee is a mid-wing, high-speed, subsonic, remotely controlled guided missile. It is propelled by a single turbojet engine. During flight, the missile is controlled from the ground. It is recovered by a parachute system. The mission of the Firebee is to support the evaluation of air defense artillery missile and gun systems. It has the following characteristics.

Speed – 500 knots
Altitude – 100 to 61,000 feet
Endurance – 90 minutes

Figure B-49. MQM-34D Firebee

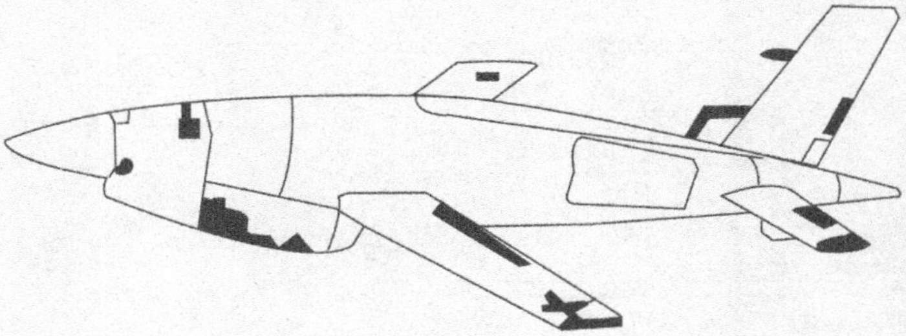

NOTE: The BQM-34 is the most powerful subscale target available.
Speed: 220 to 550 knots
Altitude: 100 to 55,000 feet
Flight time: 60 minutes
Payload: Internal 380 pounds, external 600 pounds each wing

**Figure B-50. BQM-34 Firebee**

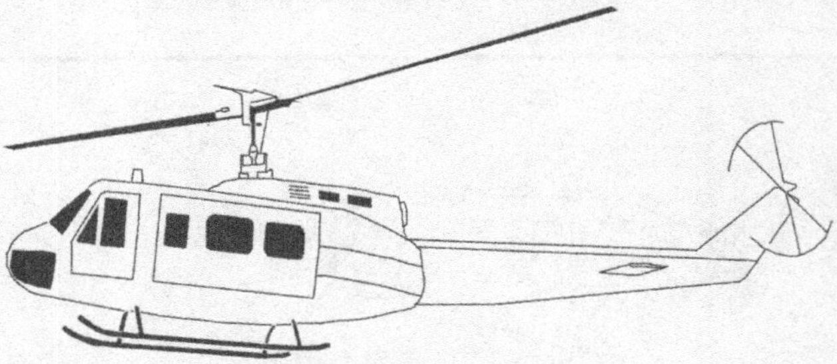

NOTE: The QUH-1 Huey is a full-scale aircraft that can perform numerous flight profiles from low altitude hover to normal flight altitudes.
Speed: Hover to 80 knots
Altitude: 10,000 feet
Flight time: 90 minutes
Payload: 1,000 pounds
Limited nap-of-the-earth

**Figure B-51. QUH-1 Huey**

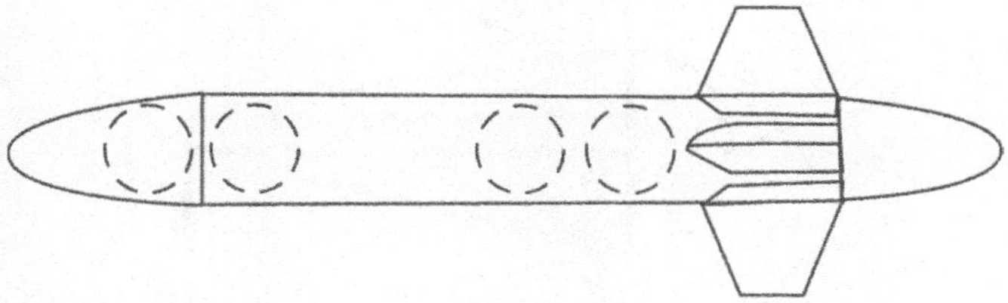

NOTE: The radar cross section (RCS) of the Patriot Omni directional Training Aerial (POTA)-Tow is designed to meet the requirements for Patriot live-fire training.

RCS: 3 to 5 square meters in C and S band
6,000 feet of 0.065-inch towline
Miss distance scoring available
Towed by MQM-107, T-38, or F-16

Figure B-52. POTA-Tow

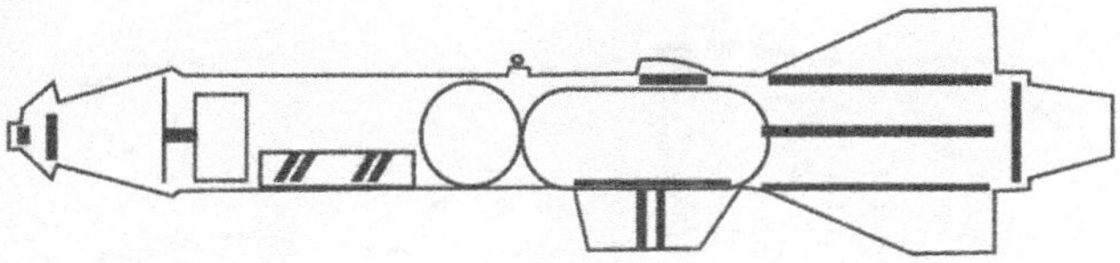

NOTE: This infrared threat target has both the visual and IR augmentation adequate for STINGER.

IR: +140 Watts/Sr in 3 to 5 micron band
600 feet of 0.064-inch steel towline
Miss distance scoring available
Towed by MQM-107 or T-38

Figure B-53. Infrared towed target

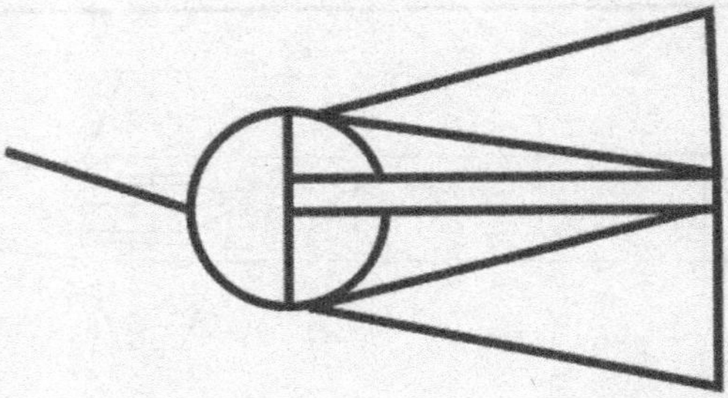

**NOTES: The low RCS tow sphere target can be towed by a MQM-107 or a T-38 from an 8,000 foot kevlar cable.**

**RCS: Less than 0.1 square meter in the C-band.**

**Figure B-54. Low radar cross section tow sphere target**

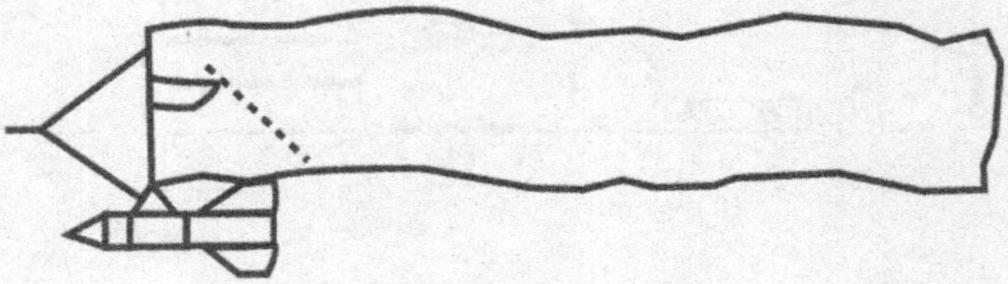

**NOTES: The snatch banner consists of a 2 foot by 12 foot banner display with a corner reflector and bullet counter.**

**RCS: One square meter in X-band.**
**600 feet of 3/8-inch diameter nylon tow rope.**
**Bullet counting scoring available.**
**Towed by MQM-107.**

**Figure B-55. Snatch banner target**

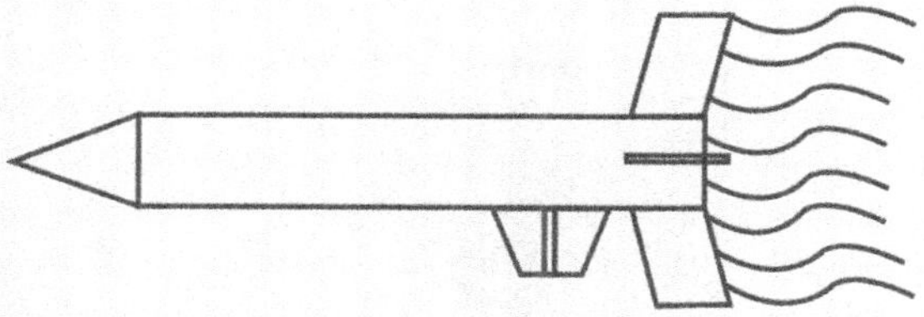

**NOTE: RCS is one square meter in the X-band. It has 600 feet of 0.064-inch diameter steel towline, bullet counting scoring available and is towed by the MQM-107.**

**Figure B-56. Aerial gunnery tow target**

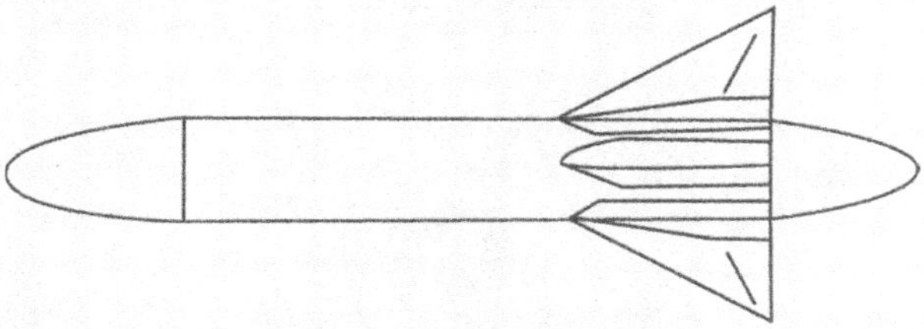

**NOTE: An X-band radar augmented target towed on an 8,000 foot steel monofilament cable under the MQM-107. RCS: 8 square meters in X-band.**

**Figure B-57. TRX-4A Towbee**

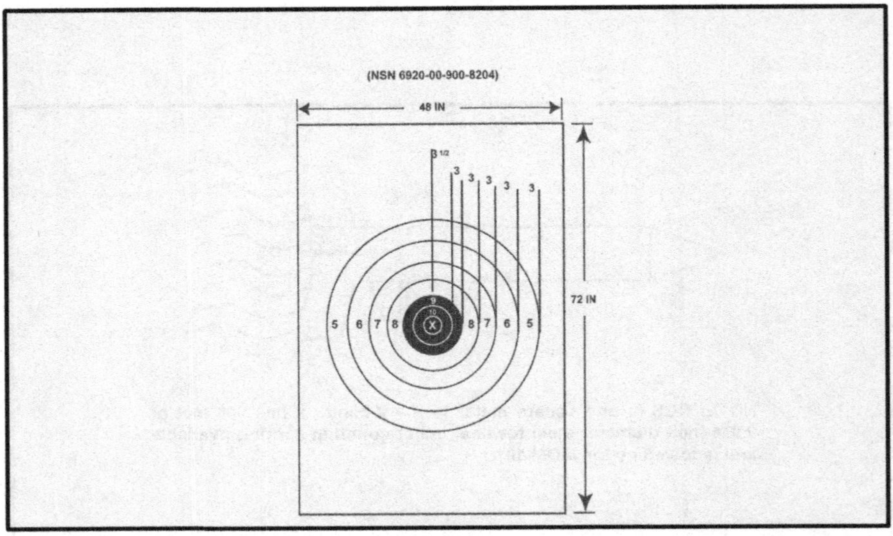


**Figure B-58. Army rifle target A, 200-300 yards**

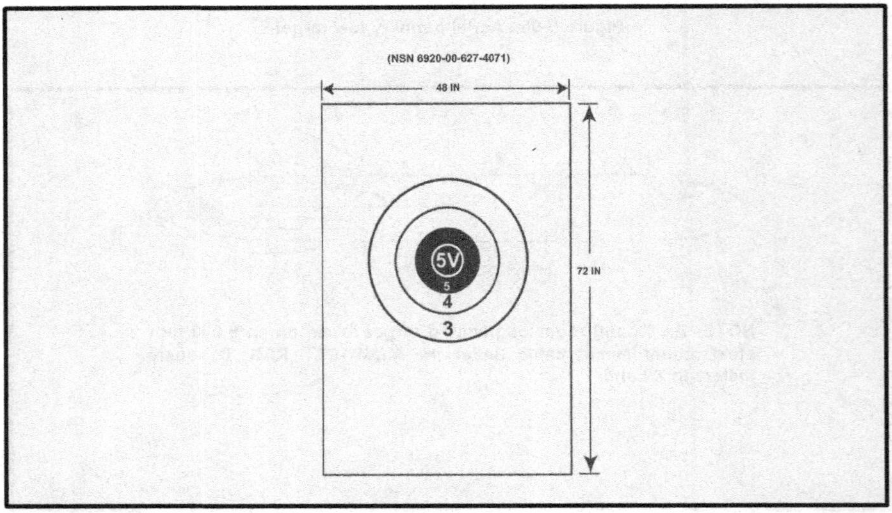


**Figure B-59. Navy and Marine Corps rifle target A, 200-300 yards**

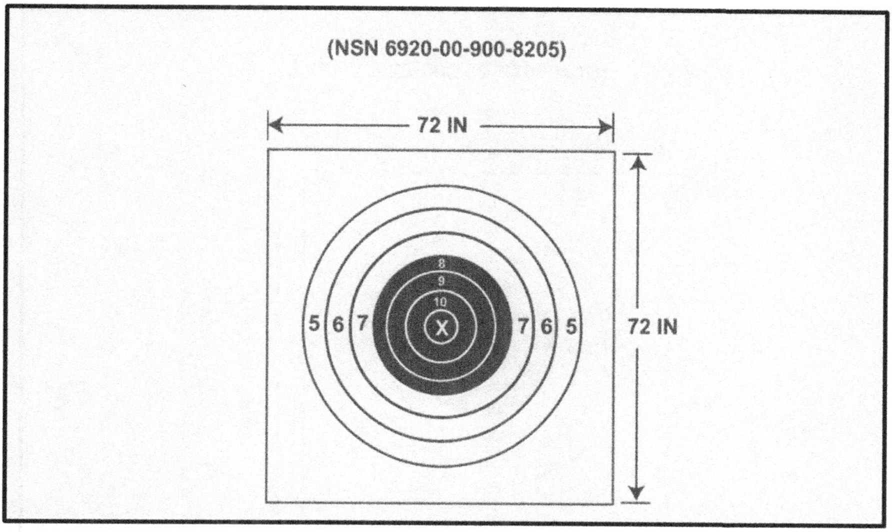


Figure B-60. Army rifle target B, 600 yards

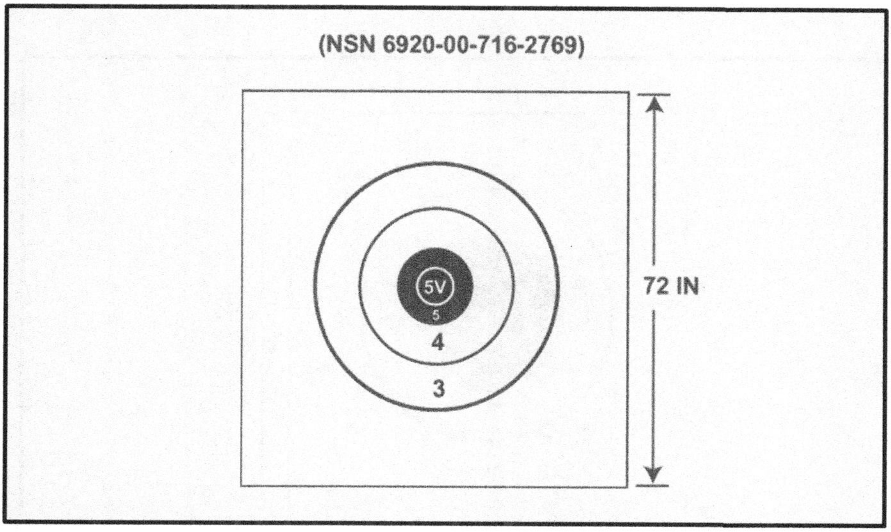


Figure B-61. Navy and Marine Corps rifle target B, 600 yards

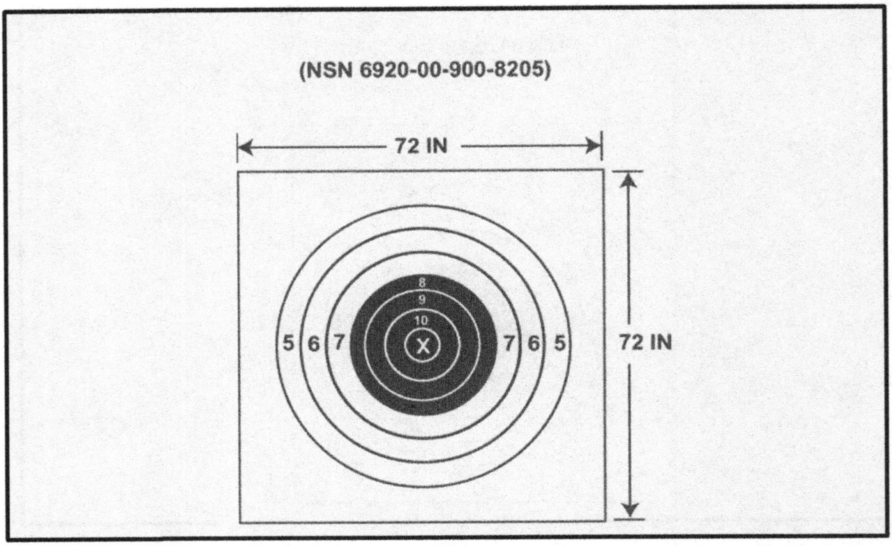


**Figure B-62. Army rifle target C, 1000 yards**

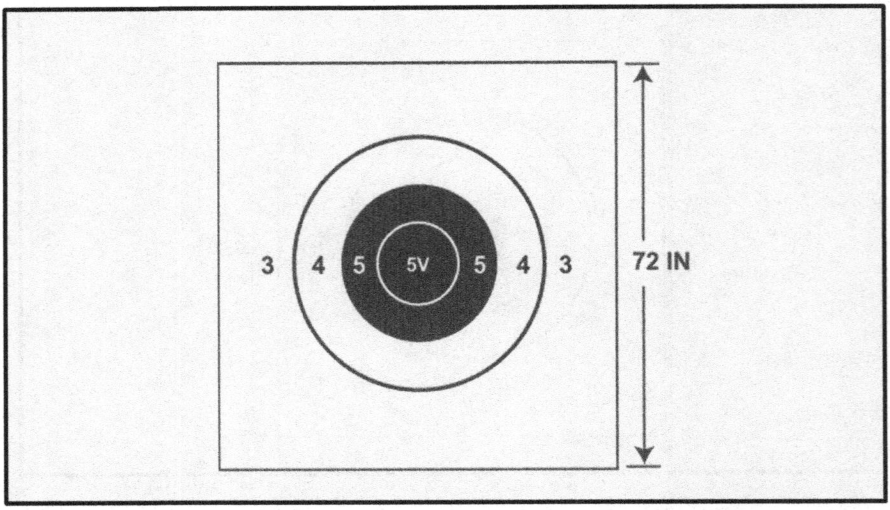


**Figure B-63. Navy and Marine Corps rifle target C, 1000 yards**

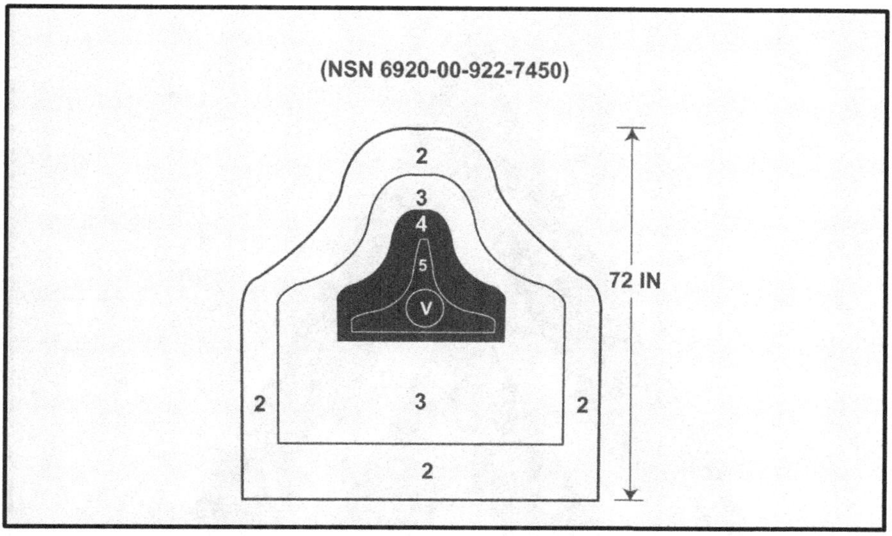


Figure B-64. Navy and Marine Corps target D, prone

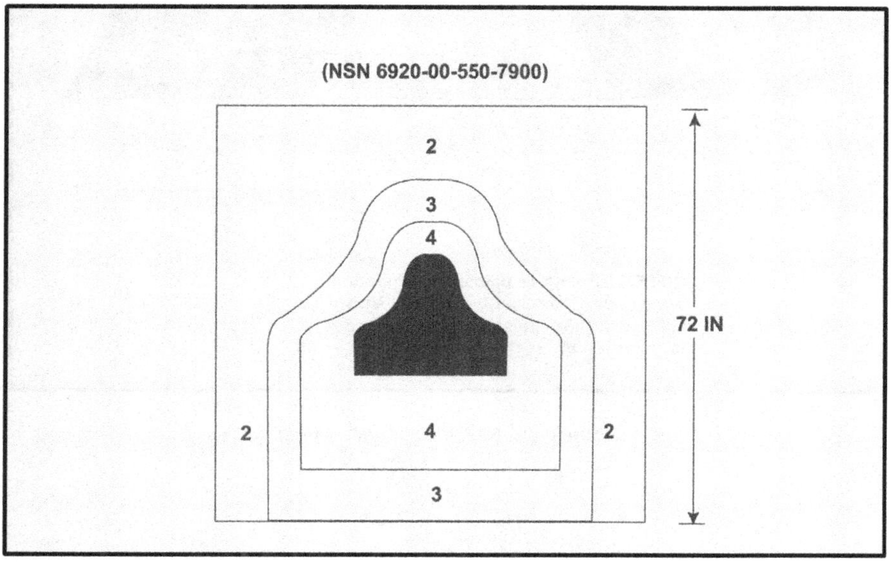


Figure B-65. Navy and Marine Corps target D, prone

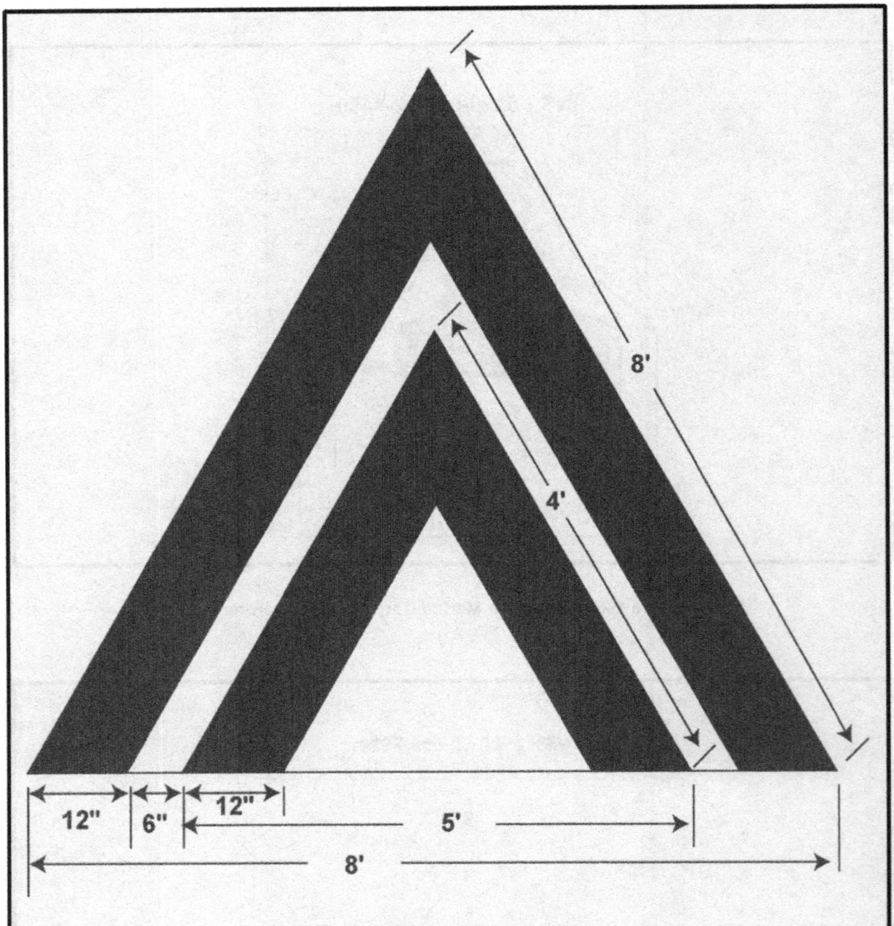


Figure B-66. Range outer limit marker

RANGE LANE MARKER UP TO 300 METERS

**NOTE: Mark even numbered lanes with white numerals on black background. Mark odd numbered lanes with black numerals on a white background.**

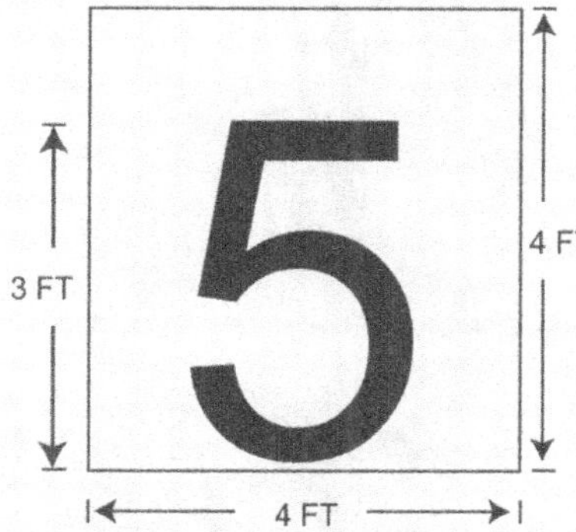


RANGE LANE MARKER UP TO 300 METERS

**Figure B-67. Range lane markers**

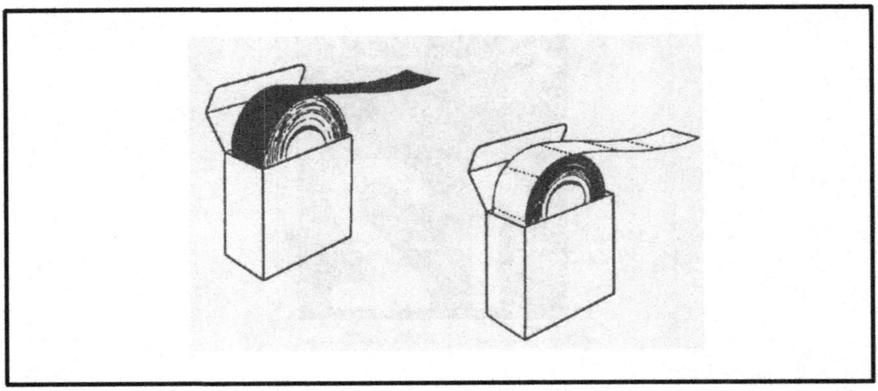

Figure B-68. Target pasters

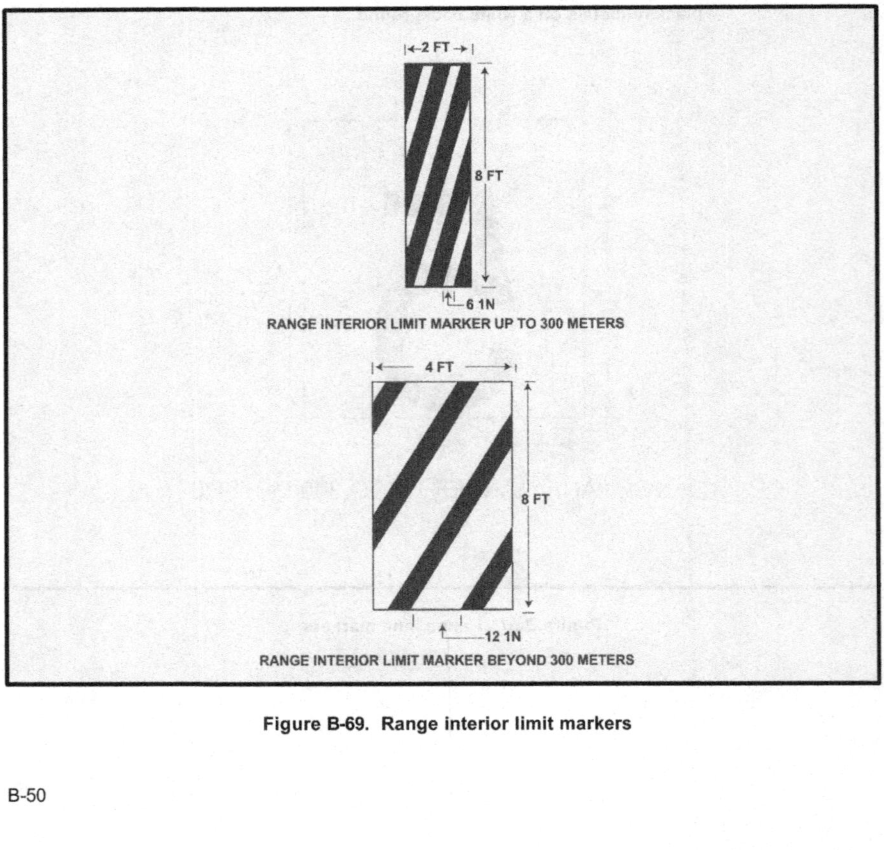


Figure B-69. Range interior limit markers

# Appendix C

## C-1 Live-Fire Checklist

Officers-in-charge, range safety officers, laser safety officers and unit trainers use Tables C-1 through C-5 checklist to plan and conduct live-fire training. The five-part checklist is a guide; it must be supplemented by AR/DA PAM 385-63, and appropriate TMs and FMs, Military Handbook 828A, local range regulations, special weapons requirements, and unit SOPs.

| Table C-1 . Live Fire Part I Preplanning Checklist | | |
|---|---|---|
| **A.** | **Mission analysis** | |
| | 1. | Units firing on the range<br>Units: ______________________<br>Number of soldiers: ______________________ |
| | 2. | Weapons and course/table to be fired<br>Weapons: ______________________<br>Course/table: ______________________ |
| | 3. | Range requested<br>Range name/number: ______________________ |
| | 4. | Date of scheduled operation<br>Date: ______________________ |
| | 5. | Ammunition requisition<br>Type/quantities required: ______________________<br>Date requested: ______________________ |
| | 6. | Attend range-control safety briefing<br>Date: ______________________ |
| | 7. | Adhere to laser safety considerations in Military Handbook 828A<br>______________________ |
| | 8. | Targets ordered: ______________________<br>Types: ______________________<br>Date: ______________________ |

**Table C-1 . Live Fire Part I Preplanning Checklist (cont)**

| | | | Y | N | NA |
|---|---|---|---|---|---|
| **B.** | **Administrative requirements** | | **Y** | **N** | **NA** |
| | 1. | Sufficient ammunition has been requested | | | |
| | 2. | Overhead fire ammunition required | | | |
| | 3. | Sufficient time has been scheduled to complete training | | | |
| | 4. | Firing periods coordinated with Range Control | | | |
| | 5. | Range-scheduling conflicts have been resolved | | | |
| **C.** | **Personnel requirements** | | **Y** | **N** | **NA** |
| | 1. | OIC | | | |
| | | Attended safety course briefing | | | |
| | | Completed unit certification | | | |
| | 2. | RSO | | | |
| | | Attended safety course briefing | | | |
| | | Completed unit certification | | | |
| | 3. | LRSO | | | |
| | | Attended safety course briefing | | | |
| | | Completed unit certification | | | |
| | 4. | Assistant safety officers | | | |
| | 5. | Medical support | | | |
| | 6. | Ammunition detail | | | |
| | 7. | Assistant instructors | | | |
| | 8. | Range guards | | | |
| | 9. | Range/target operator | | | |
| | 10. | Maintenance personnel | | | |
| | 11. | Target detail | | | |
| **D.** | **Equipment requirements** | | **Y** | **N** | **NA** |
| | 1. | Range packet/clearance form received | | | |
| | 2. | Safety fan diagram/range overlay on hand | | | |
| | | Firing points | | | |
| | | Firing lanes | | | |
| | | Firing boxes | | | |
| | | Target locations | | | |
| | 3. | Radios | | | |
| | 4. | Range flag and light (night firing) | | | |
| | 5. | Vehicle flag sets/lights | | | |
| | 6. | Safety paddles | | | |
| | 7. | Ambulance or designated vehicle | | | |

**Table C-1 . Live Fire Part I Preplanning Checklist (cont)**

| | | | | | |
|---|---|---|---|---|---|
| | | Litter | | | |
| | | Backboard | | | |
| | | First-aid bag | | | |
| | | Other equipment prescribed in local guidance | | | |
| | 8. | Earplugs | | | |
| | 9. | Master score sheet | | | |
| | 10. | Scorecards | | | |
| | 11. | Armorer's tool kit | | | |
| | 12. | Weapons-cleaning equipment | | | |
| | 13. | Fire extinguishers | | | |
| | 14. | Repair parts/spare weapons | | | |
| | 15. | Training publications, range and safety publications | | | |
| | 16. | Report folder | | | |
| | 17. | Lifesaving procedures | | | |
| | 18. | Sanitary and cleanup supplies | | | |

| **Table C-2. Live Fire Part II Range Occupation** | | | | | |
|---|---|---|---|---|---|
| **A.** | **Establish communications with Range Control** | | **Y** | **N** | **NA** |
| | 1. | Establish FM radio range control firing net | | | |
| | 2. | Establish FM radio range control administrative net | | | |
| | 3. | Establish Radio or telephone range control back-up firing net | | | |
| **B.** | **Medical support is present** | | **Y** | **N** | **NA** |
| | 1. | Medics have communications with treatment facility | | | |
| | 2. | Medics have strip map from range to treatment facility | | | |
| | 3. | Medics (off-site) have communications with range | | | |
| | 4. | Medics (off-site) have strip map to range | | | |
| | 5. | Medics, in coordination with range OIC, select and clear air evacuation site near the range | | | |
| | 6. | Target locations | | | |
| **C.** | **Prepare designated areas** | | **Y** | **N** | **NA** |
| | 1. | Ammunition point | | | |
| | 2. | Medical station | | | |
| | 3. | Concurrent training | | | |
| | 4. | Parking | | | |
| | 5. | Armorer | | | |
| | 6. | Water point | | | |
| | 7. | Mess | | | |

**Table C-2. Live Fire Part II Range Occupation (cont)**

| | | Y | N | NA |
|---|---|---|---|---|
| **D.** | **Range OIC brief** | | | |
| | 1. RSO | | | |
| | 2. LRSO (if required) | | | |
| | 3. Safety assistants | | | |
| | 4. Medical NCOIC | | | |
| | 5. Assistant instructors | | | |
| | 6. Vehicle commanders | | | |
| **E.** | **Range guards briefed and posted** | | | |
| **F.** | **Inspect range to ensure targetry is present and operational** | | | |
| **G.** | **Verify impact area clear of unauthorized personnel** | | | |
| **H.** | **Raise range flag** | | | |
| **I.** | **Check ammunition (No live-fire ammunition on non-firing range)** | | | |
| **J.** | **Receive firing units** | | | |
| **K.** | **Conduct safety checks on weapons** | | | |
| | 1. Dud orientation | | | |
| | 2. Noise hazard briefing | | | |
| | 3. Misfire procedures | | | |
| **M.** | **Organize personnel into firing orders (keep unit integrity, if possible)** | | | |
| **N.** | **For tanks and BFVs, inspect DA Form 2408-4 (Weapon Record Data) for each main gun to be fired** | | | |
| **O.** | **Request clearance from Range Control to commence firing** | | | |

| **Table C-3. Live Fire Part III Firing Operations** | | Y | N | NA |
|---|---|---|---|---|
| **A.** | **Communications maintained with Range Control** | | | |
| **B.** | **Ammunition accountability maintained** | | | |
| **C.** | **Personnel accountability maintained** | | | |
| **D.** | **Guards on duty/alert** | | | |
| **E.** | **Earplugs in use** | | | |
| **F.** | **Weapons cleared before departing firing line (rod on small-arms ranges)** | | | |
| **G.** | **Surface danger zone is monitored to ensure it remains clear** | | | |
| **H.** | **Laser danger zones are monitored for unsafe acts or improper safety equipment** | | | |
| **I.** | **Weapon systems are cleared and checked during temporary suspensions of firing** | | | |
| **J.** | **Cease-fire is called when unsafe act is observed or reported or when communications with Range Control is lost** | | | |
| **K.** | **Handle misfires, hangfires, or cookoffs in accordance with technical manuals** | | | |

| Table C-4. Post-firing Operations | | Y | N | NA |
|---|---|---|---|---|
| A. | Weapons cleared before leaving range | | | |
| B. | Conduct brass/ammunition check | | | |
| C. | Close range down in accordance with SOP | | | |
| D. | Firing status of range or vehicles reported to Range Control | | | |
| E. | Conduct police of range | | | |
| F. | Perform maintenance tasks as required by range SOP | | | |
| G. | Request range clearance and shut-down from Range Control | | | |
| H. | Secure range | | | |

Laser devices may be used only on those ranges that the installation commander has approved and established for such use. The following tables and checklists are furnished as guides for planning and conducting laser operations. Cautions, warnings, and hazards of laser systems must be strictly followed and enforced. Military Handbook 828A should be used as a reference and guide.

| Table C-5. Laser Operations | | Y | N | NA |
|---|---|---|---|---|
| A. | Verify survey of proposed lasing and target area | | | |
| B. | Right and left laser safety limit stakes designate right and left limits of lasing at local training areas | | | |
| C. | Warning signs and barricades posted to prevent unauthorized entry | | | |
| D. | Warning signs posted at entrance to the range | | | |
| E. | Verify impact area clear of unauthorized personnel | | | |
| F. | Sweep range area to remove reflective material before lasing | | | |
| | 1. Target materials are non-reflective surfaces – cardboard, wood, or lusterless metal | | | |
| | 2. Recommended target areas are free of reflective surfaces (glossy foliage and raindrops are not considered reflective surfaces) | | | |
| | 3. Laser devices are not lased at specular (mirror like surfaces) | | | |
| G. | Unprotected personnel are not exposed to either the direct beam or the beam reflected from a specular surface | | | |
| H. | Personnel within the laser safety danger zone are wearing protective eyewear | | | |
| J. | Appropriate laser safety filters are placed in daylight optical devices used to observe targets during lasing | | | |
| K. | Recommended target areas are free of calm, smooth water, and clean ice | | | |
| L. | Laser safety orientation is provided to personnel who work with or use lasers | | | |

# Appendix D
# Range Diagrams

## D-1. Instrumented RangesFunctional Description

The Instrumented Range program is composed of the Digital Multi-Purpose Range Complex (DMPRC), the Digital Multi-Purpose Training Range (DMPTR), and the Battle Area Complex (BAX). Due to the commonality of the solutions among these various types of ranges, this consolidation represents the most economical and practical use of development assets. The individual functions of these three ranges are described below:

## D-2. Digital Multi-Purpose Range Complex (DMPRC):

a. Supports platoon level collective training for all gunnery tables for Abrams, Bradley, (and all dismounted infantry) and aviation.
b. Supports individual and crew qualification training.
c. Uses thermal targets, night illumination devices, hostile-fire, target-kill, and visual flash simulators.

## D-3. Digital Multi-Purpose Training Range (DMPTR):

a. Designed as one of two variants: Current Force or Stryker vehicle.
b. Supports crew qualification for up to table VIII.
c. Supports dismounted infantry squad training independently or with supporting vehicles.
d. Supports crew/squad level training and individual and crew qualifications.
e. Uses thermal targets, night illumination devices, hostile-fire, target-kill, and visual flash simulators.
f. Serves as feeder range for DMPRC, BAX and other collective training facilities.

## D-4. Battle Area Complex (BAX):

a. Supports infantry-centric platoon/company CALFEX with MGS crew qualification overlay.
b. Incorporates complex terrain, bunkers, trenches, live fire villages, shoot houses and gunnery type facades.
c. Provides engineer area, add-on urban and mortar impact areas.
d. Uses thermal targets, night illumination devices, hostile-fire, target-kill, and visual flash simulators.

This "product line" of instrumented ranges provides enhanced training data collection for After Action Reviews (AARs), utilization of interfaces compliant with the Common Training Instrumentation Architecture (CTIA), use of advanced targetry to represent realistic threats, and the stimulation of digitized force systems. System and component commonality of the product line will offer a significant future payback in the development and fielding of the instrumented ranges.

## D-5. Major Components:

The major components (common to all three instrumented range types) consist of the following:

a. Range Operations Center (ROC)
b. Player Unit (PU) Instrumentation
c. Player Instrumentation Radio (PIR)
d. Audio/Video Recording and Control
e. After Action Review (AAR)
f. Range Targetry
g. Battlefield Effects System (BES)

*This drawing shows an example layout of an instrumented range:*

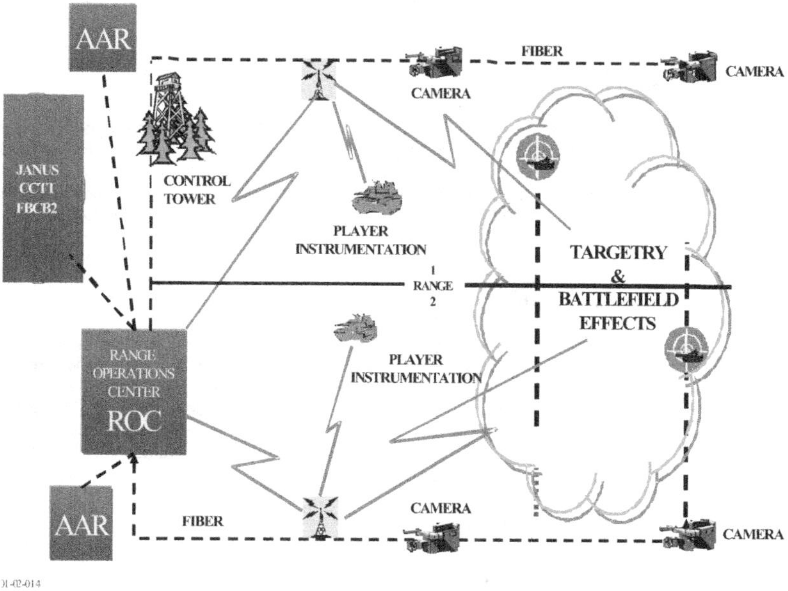


## D-6. Range Operations Center (ROC)

a. The ROC is the heart of the instrumented range. It provides an Integrated Management System (IMS) to control all functions from a single workstation. As many workstations as required could be connected to the network to provide control or access to data. The ROC architecture is predicated on the philosophy that all data are stored in a relational database. This includes real-time range data, manually entered user data, data commands, user roles and simulation, targetry data, and overall results data. The ROC will use a platform-independent relational database.

b. The ROC will have multiple functions. It will serve as the central control point for all data collected and for control of all the components of the range facility. The ROC will provide the capability to control targetry, the interface for target data collection, and the capability to populate the range database with data collected from targets. The database will not only provide central processing of targetry information, but it will also collect information from the player instrumentation, video equipment, and the audio components. A High Level Architecture (HLA) interface will be provided at the ROC for external simulations (such as CCTT or Janus). A situational display will provide ROC operators with a view of the player unit (PU) status and range awareness for safety and control of the exercise. The ROC will also provide an overall mission scenario authoring capability to synchronize all Digital Range components.

## D-7. Player Unit (PU) Instrumentation

The PU Instrumentation, comprised of all the communications, interfaces and other electronic packages mounted on the host platforms (target, vehicle, aircraft, or crew-served weapon) to be utilized in the training scenarios. The PU interfaces with the host weapons which records and forwards information to the ROC. Data that is forwarded includes Real Time Casualty Assessments (RTCA) results, Force XXI Battle Command Brigade and Below (FBCB2) traffic, through-sight video, position location, weapon system data bus parameters, and simulated live-fire pairing.

**D-8. Player Instrumentation Radio (PIR)**

The PIR is a bi-directional communication link between the ROC and PU instrumentation in support of near-real-time monitoring and controlling of the instrumented ranges live platform functions. Relayed data includes vehicle status, crew performance, AAR presentations, and range safety information. The PIR receives administrative commands from the ROC via a Network Control Processor (NCP) interface. The NCP ensures messages are received correctly, logged, and transmitted over the ROC LAN for near-real-time use.

**D-9. Audio/Video Recording and Control**

The audio/video recording system will enable users to easily tag, search, retrieve and playback key segments through a distributive system throughout the ROC. The audio recording system will record tactical radio transmissions, observer/controller (OC) radio systems and telephone systems. Through - sight video and range cameras will record activities in the exercise. The video and audio recording components will be integrated into a comprehensive system that provides an end-to-end digital asset management solution with the ability to acquire, retrieve, transmit, and display digital assets. Video feeds can be used for after-action reviews as well as for range safety monitoring.

**D-10. After Action Review (AAR)**

The AAR creation process starts with the execution of the exercise. Since there is a real time monitoring capability of the exercise, any operator can tag informational events, FBCB2 messages, or audio/video, for further analysis. The database allows data to be easily searched and retrieved based on time tags, or ad-hoc queries. Therefore, statistical and event data can be quickly formulated into a presentation, or slides that can be automatically created from saved templates. For situational awareness, the video and audio assets can be synchronized and played back in unison as part of an AAR. Users will be able to export situational displays along with multimedia assets into slide presentations, web pages, printed material, or CD/DVD. The ability to export to CD/DVD will make AAR products available for a take-home package to be provided to training units for at-home reviews.

**D-11. Range Targetry**

Targets will include moving and stationary vehicle and personnel targets. The range system will provide target controls and lifters; the actual targets will be procured separately. These targets will interface to the ROC to provide status, position and control. The targetry component will detect and score shots taken against each target. This data will be transmitted back and stored at the ROC database for use in the AAR. Targetry will be able to interact with the Tactical Engagement Simulation (TES). This will allow shootback capabilities.

**D-12. Battlefield Effects System (BES)**

Battlefield effects on the instrumented ranges will include, but are not limited to, smoke, sound, and pyrotechnics (explosions). These effects will be controlled from the ROC.

## Figure D-1. FCC 17801 RIFLE/MACHINE GUN ZERO RANGE

This range is used to train individual soldiers on the skills necessary to align the sights and practice basic marksmanship techniques against stationary targets. The range is designed for training Shot-Grouping and Zeroing exercises with the M16 and M4 series rifles as well as crew served machine guns.

Primary features include:
- 32 Target frames at 25m
- 16 Target frames at 10m
- 32 Foxholes

This range requires no automation. All targets are fixed at 25 meters from the firing line for M16/M4 and fixed at 10 meters for machine gun.

Associated Range Operations and Control facilities:
- Standard Small Arms ROCA Facilities

**Requirement Document:** FM 3-22.9, FM 23-14, FM 23-65,
**Additional Information:** None

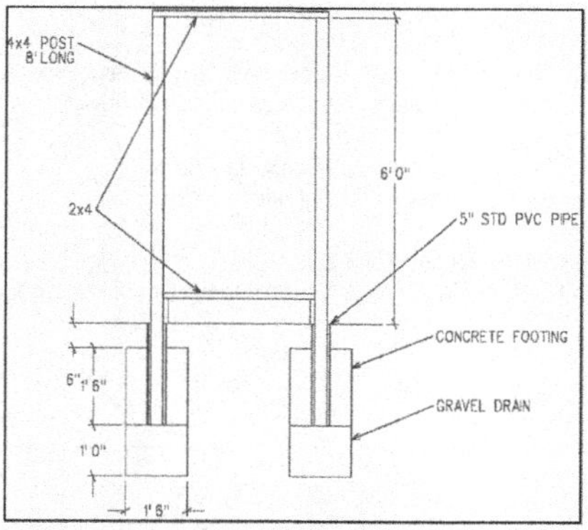


TARGET BOOT DETAIL

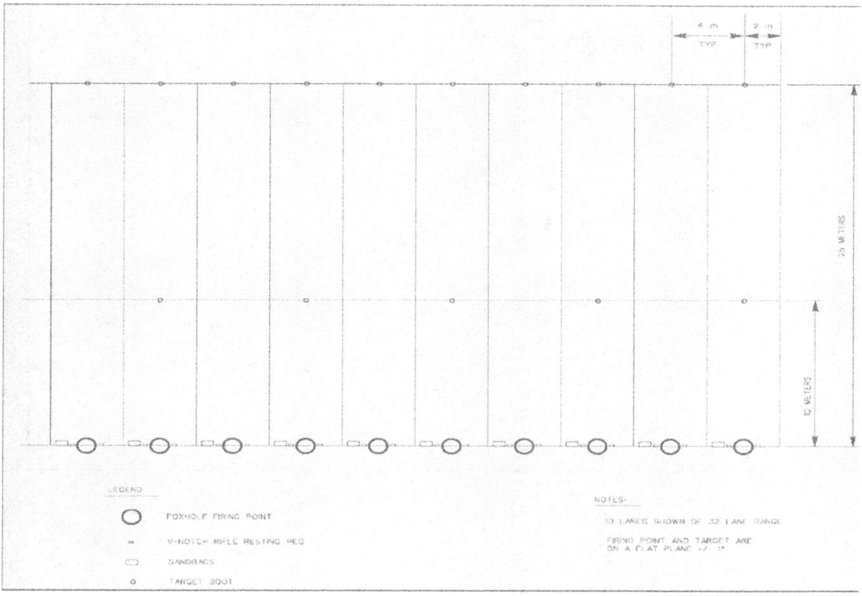

## Figure D-2. FCC 17803 AUTOMATED FIELD FIRE (AFF) RANGE

This range is used to train and familiarize soldiers on the skills necessary to identify, engage and hit stationary infantry targets with the M16 & M4 rifles.

Primary features include:
- 96 Stationary infantry targets
- 32 Foxhole positions

All targets are fully automated and the event specific target scenario is computer driven and scored from the range operations center. The range operating system is fully capable of providing immediate performance feedback to the using participants.

Associated Range Operations and Control facilities:
- Standard Small Arms ROCA Facilities

**Requirement Document:** FM 3-22.9
**Additional Information:** This range can be used for automatic rifle practice.

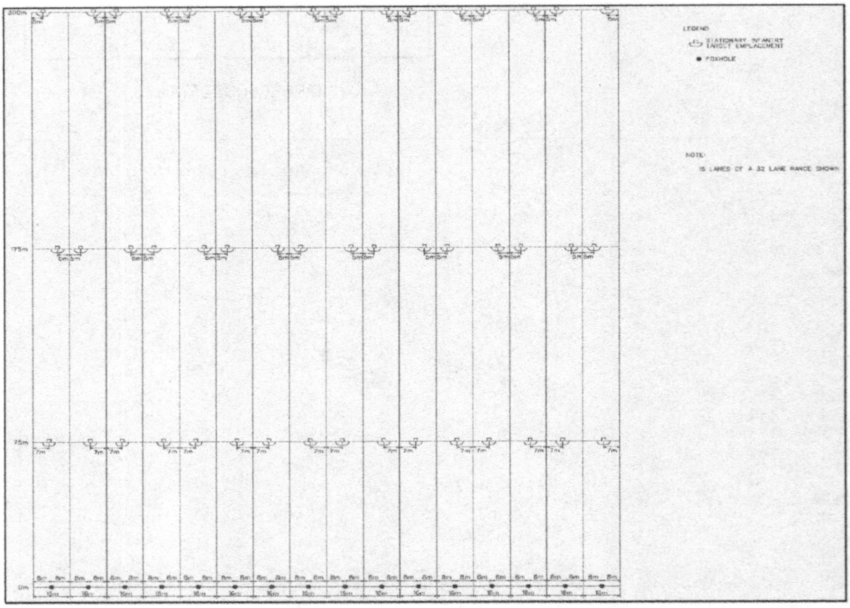

## Figure D-3. FCC 17805 AUTOMATED RECORD FIRE (ARF) RANGE

This range is used to train and test individual soldiers on the skills necessary to identify, engage and defeat stationary infantry targets for day/night qualification requirements with the M16 & M4 rifles.

Primary features include:
- 112 Stationary infantry targets
- 16 Foxholes

All targets are fully automated and the event specific target scenario is computer driven and scored from the range operations center. The range operating system is fully capable of providing immediate performance feedback to the using participants.

Associated Range Operations and Control facilities:
- Standard Small Arms ROCA Facilities

**Requirement Document:** FM 3-22.9
**Additional Information:** Night firing is accomplished from the night-fire line and baseline, firing at the 50m targets. Replace one of the 50m F-type silhouettes with an E-type silhouette. Low light illumination capability is required in both of the 50m target emplacements. Engage the F-type from the 25m night-firing line, and the E-type from the baseline.

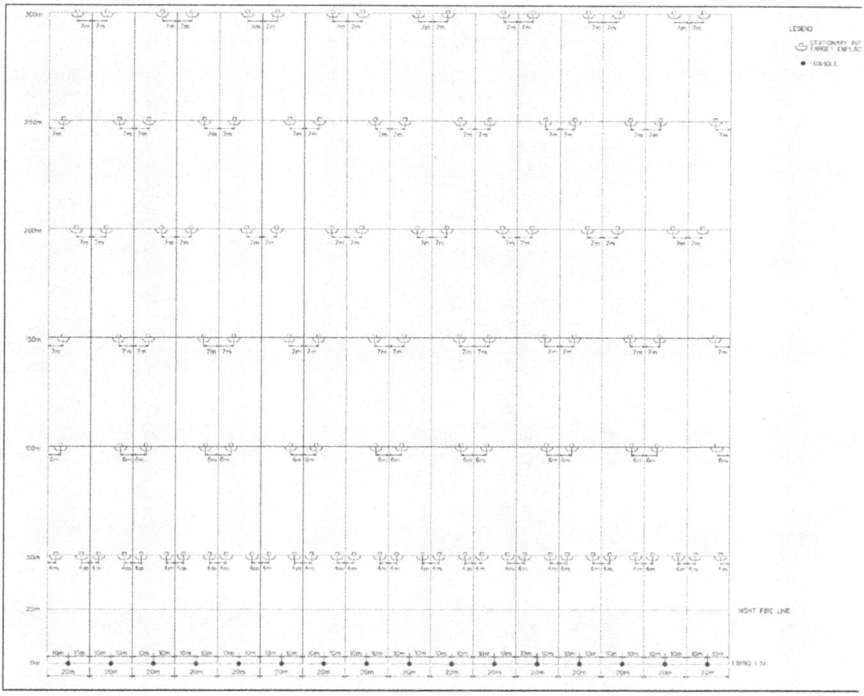

## Figure D-4. FCC 17806 MODIFIED RECORD FIRE (MRF) RANGE

This range is used to train and test individual soldiers on the skills necessary to identify, engage and defeat stationary infantry targets for day/night qualification requirements with the M16 & M4 rifles. This range combines the capabilities of Automated Field Fire (17803), Automated Record Fire (17805), and the Automated Night Fire (17808) to reduce land and maintenance requirements and increase efficiencies.

Primary features include:
- 144 Stationary infantry targets
- 16 Foxholes

All targets are fully automated and the event specific target scenario is computer driven and scored from the range operations center. The range operating system is fully capable of providing immediate performance feedback to the using participants.

Associated Range Operations and Control facilities:
- Standard Small Arms ROCA Facilities

**Requirement Document:** FM 3-22.9

**Additional Information:** Night firing accomplished from night-fire line and baseline, firing at the 50m targets. Replace one of the 50m F-type silhouettes with an E-type silhouette. Low light illumination capability is required in both of the 50m target emplacements. Engage the F-type from the 25m night-firing line, and the E-type from the baseline.

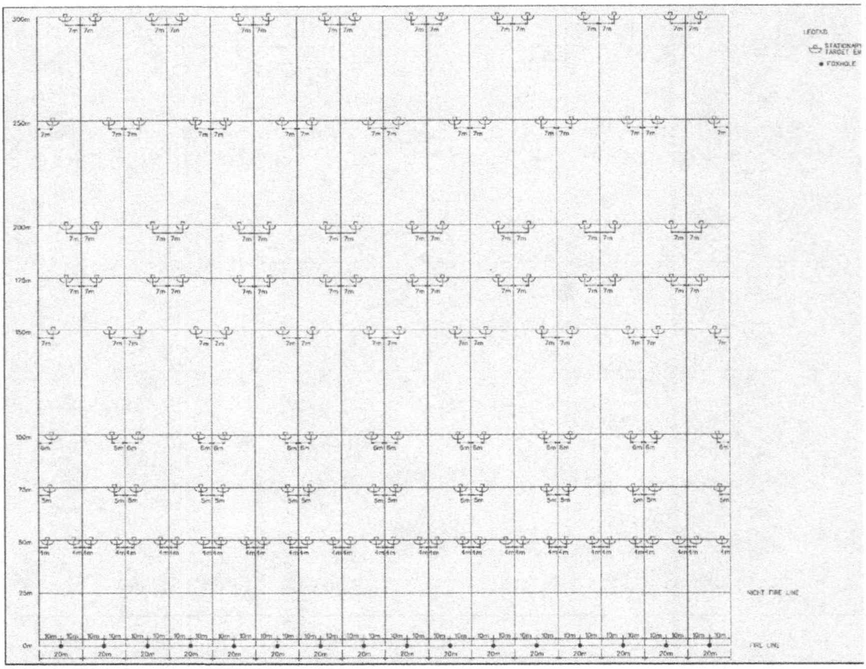

**Figure D-5. FCC 17809 QUALIFICATION TRAINING RANGE (QTR)**

This range is used to train and test soldiers on the skills necessary to detect, identify, engage and defeat stationary and moving infantry targets in a tactical array with their prescribed weapons. This range combines the capabilities of Modified Record Fire Range (17806), Automated Sniper Field Fire Range (17812), Combat Pistol/Military Police Firearms Qualification Course (17822), and the Multipurpose Machine Gun Range (17833) to centralize training and reduce land, maintenance, and unit overhead requirements.

Primary features include:
- 15 Lanes Combat Pistol Qualification
- 10 Lanes Sniper Field Fire (see notes)
- 16 Lanes Modified Record fire
- 10 Lanes Multipurpose Machine Gun
- 32 Lanes Rifle/Machine Gun Zero

- 429 Stationary infantry targets
- 20 Stationary armor targets
- 20 Moving infantry targets
- 10 Sationary infantry emplacements with 2 target mechanisms each (400 m and 700 m)

All targets are fully automated and the event specific target scenario is computer driven and scored from the range operations center. The range operating system is fully capable of providing immediate performance feedback to the using participants.

Associated Range Operations and Control facilities:
- Standard Small Arms ROCA Facilities
- Additional General Instruction Building (17120)
- Additional Ammo Breakdown Building (17129)

**Requirement Document:** FM 3-22.9, FM 3-22.10 (23-10), FM 3-22.35 (23-35), FM 19-10, FM 23-65, FM 3-22.68
**Additional Information:** This range enhances throughput capability for units with multiple weapons densities by consolidating unit efforts to operating one training facility.

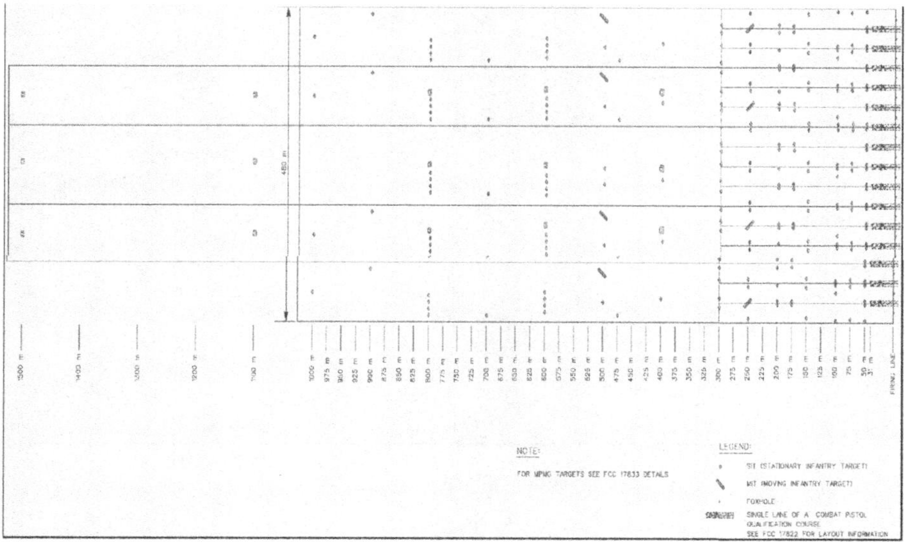

## Figure D-6. FCC 17812 AUTOMATED SNIPER FIELD FIRE RANGE

This range is used to train and test soldiers on the skills necessary to detect, identify, engage and defeat stationary and moving infantry targets in a tactical array. This range is designed to satisfy the training and qualification requirements of the M24 Sniper rifle.

Primary features include:
- 40 stationary infantry targets
- 8 Moving infantry targets
- 4 firing positions

All targets are fully automated and the event specific target scenario is computer driven and scored from the range operations center. The range operating system is fully capable of providing immediate performance feedback to the using participants.

Associated Range Operations and Control facilities:
- Standard Small Arms ROCA Facilities

**Requirement Document:** FM 3-22.10 (23-10)
**Additional Information:** Natural vegeta tion is required in the target area to provide realistic natural obstacles for the sniper to negotiate.

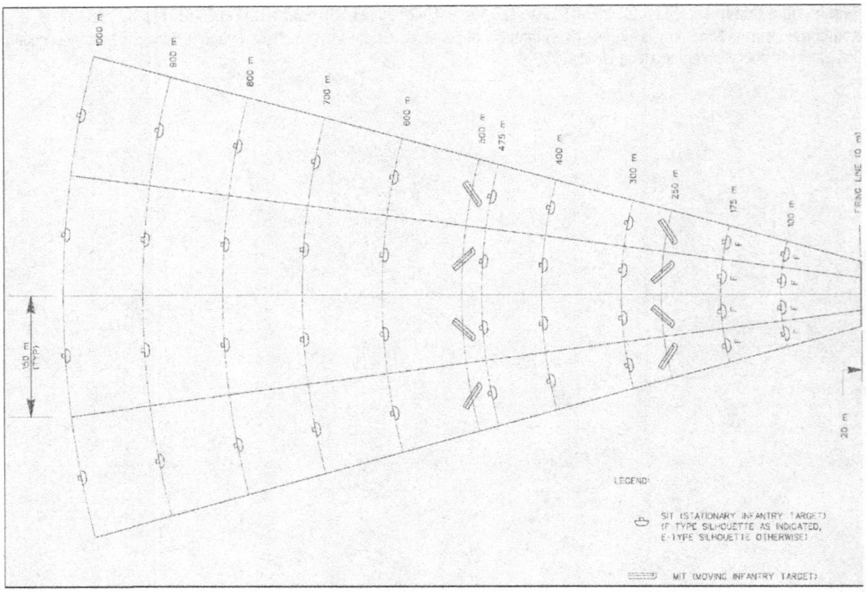


FCC 17812 AUTOMATED SNIPER FIELD FIRE RANGE

## Figure D-7. FCC 17816 BAYONET ASSAULT COURSE (BAC)

This facility is used to train individual soldiers in assault techniques using a rifle and bayonet against dummy silhouettes as well as negotiating various obstacles.

Primary features include:
- 9 Lanes
- 72 Dummy silhouettes

This facility requires no automation.

Associated Range Operations and Control facilities:
- Operations/Storage Building (17122)
- Latrine (73075)

**Requirement Document:** FM 3-25.150
**Additional Information:** The number of lanes can vary due available terrain and the number of soldiers that must complete the course.

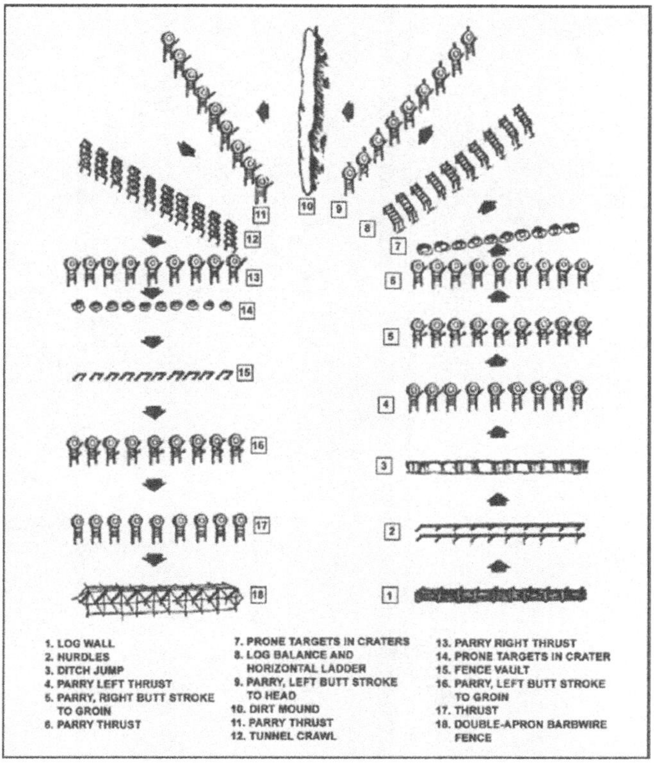

## Figure D-8. FCC 17822 AUTOMATED COMBAT PISTOL/MILITARY POLICE FIREARMS QUALIFICATION COURSE (CP/MPQC)

This range is used to train and test soldiers on the skills necessary to detect, identify, engage and defeat stationary targets in a tactical array. The complex satisfies the training and qualification requirements of the 9mm, .38 caliber, and .45 caliber pistols.

Primary features include:

- 105 Stationary infantry targets
- 15 Firing lanes
- 15 Stationary silhouettes (booted, MP Qual)

All CPQC targets are fully automated and the event specific target scenario is computer driven and scored from the range operations center. The range operating system is fully capable of providing immediate performance feedback to the participants using the CPQC. MPQC silhouettes are non-automated and is manually scored.

Associated Range Operations and Control facilities:

Standard Small Arms ROCA Facilities

**Requirement Document:** CPQC: FM 3-22.35 (23-35)
MPFQC: FM1910.
**Additional Information:** None

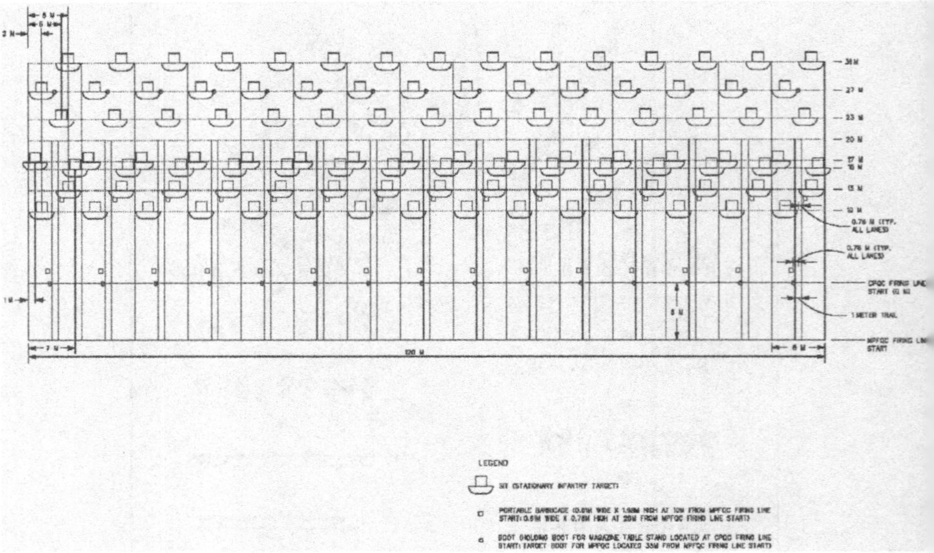

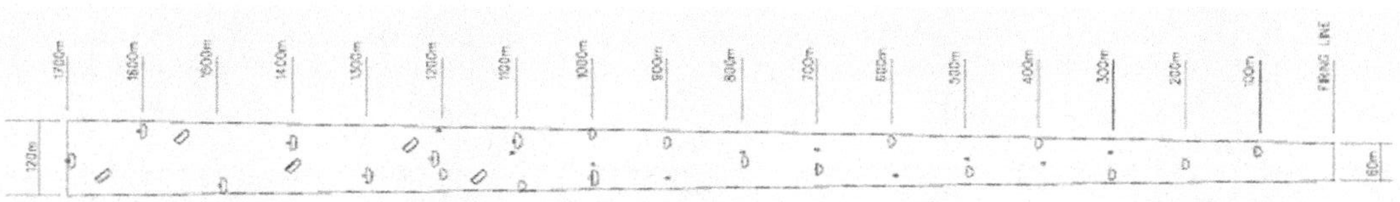


This range is used to train and test soldiers on the skills necessary to detect, identify, engage and defeat stationary and moving infantry targets in a tactical array. This range satisfies the training and qualification requirements of the heavy sniper rifle.

Primary features include:
- 12 Stationary infantry targets
- 10 Iron maidens
- 8 Stationary armor targets
- 5 Moving infantry targets

All targets, except iron maidens, are fully automated and the event specific target scenario is computer driven and scored from the range operations center. The range operating system is fully capable of providing immediate performance feedback to the using participants.

Associated Range Operations and Control facilities:
Standard Small Arms ROCA Facilities

**Requirement Document:** FM 3-22.10 (23-10).
**Additional Information:** Natural vegetation is required in the target area to provide realistic natural obstacles for the sniper to negotiate.

**Figure D-9. FCC 17829 HEAVY SNIPER RANGE**

## Figure D-10. FCC 17833 MULTIPURPOSE MACHINE GUN (MPMG) RANGE

This range is used to train and test soldiers on the skills necessary to zero M249 SAW, M60 MG, M240B MG, MK19, Sniper Weapons and M2 MG weapon systems, detect, identify, engage and defeat stationary infantry targets in a tactical array.

Primary features include:

- 180 Stationary infantry targets (SIT)
- 20 Moving Infantry Targets emplacements
- 20 Stationary armor targets
- 10 Firing lanes
- Note: Contains 20 emplacements with 2 single arm mechanisms

All targets are fully automated and the event specific target scenario is computer driven and scored from the range operations center. The range operating system is fully capable of providing immediate performance feedback to the using participants.

Associated Range Operations and Control facilities:
Standard Small Arms ROCA Facilities

**Requirement Document:** FM 23-14, FM 23-65, FM 3-22.68
**Additional Information:** Target beyond 800M will be battery powered/radio controlled.

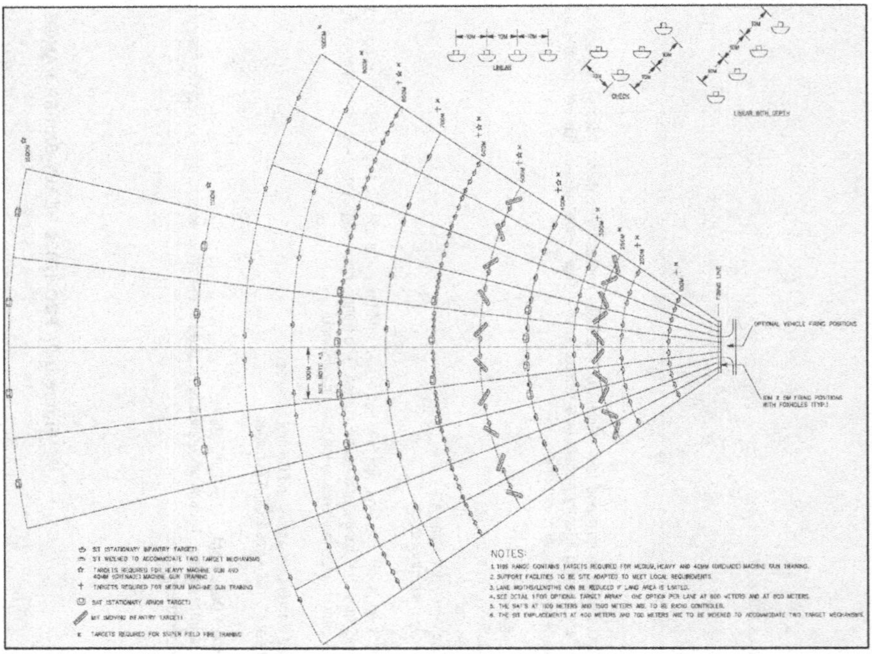

## Figure D-11. FCC 17845 ANTIARMOR TRACKING AND LIVE-FIRE

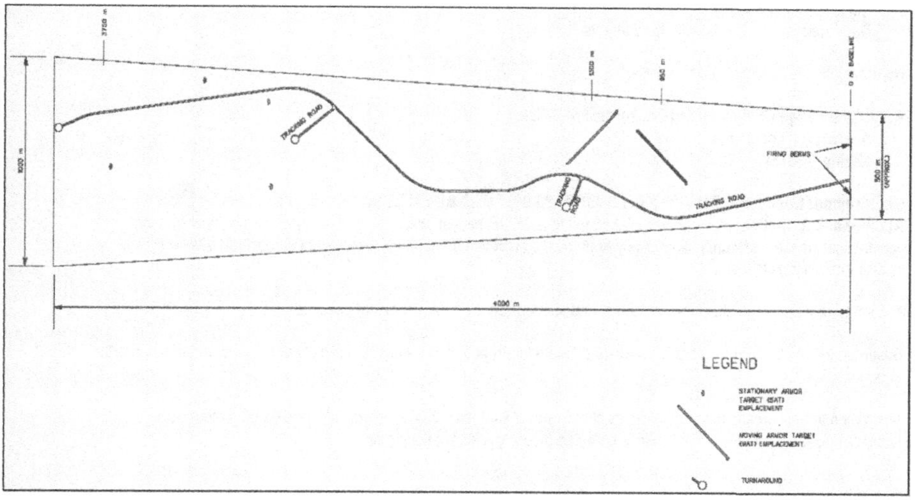


This complex is used to train and test soldiers on the skills necessary to employ weapons, identify, track, engage and defeat stationary and moving armor targets presented individually or as part of a tactical array. The complex is designed to satisfy the training and qualification requirements of medium and heavy anti-armor weapon systems.

Primary features include:
2 Moving armor targets
4 Stationary armor targets
1 Course road

All targets are fully automated, computer driven, and scored from the range operations center. The range operating system is fully capable of providing immediate performance feedback to the using participants.

Associated Range Operations and Control facilities:
Standard Armor ROCA Facilities
Except Range Operations Center, Large (17124) replaced by Range Operations Center, Small (17123);
no Ammo Breakdown Building (17129) and no Small AAR Building (17118)

**Requirement Document:** FM 3-20.8, FM 23-34, FM 23-24, FM 3-22.37, FM 3-22.999A
**Additional Information:** Location of the boresight must be coordinated with the trainer.

Gunnery tasks requiring the usage of dud producing ammunition cannot be fired on the range. Provisions for these tasks must be made in impact areas adjacent to range.

## Figure D-12. FCC 17852 MORTAR RANGE

This range is used to train mortar crews on the skills necessary to apply fire mission data, engage, and hit stationary targets in a tactical array.

Primary features include:
- Surveyed firing point(s)
- Dedicated impact area(s) with hard targets

No automation is required for this facility.

Associated Range Operations and Control facilities:
- Ammunition Loading Dock (14970)
- Latrine (73075)

**Requirement Document:** FM 7-90, FM 23-85, FM 23-90, FM 23-91, FM 23-92
**Additional Information:** This range may be next to a unit maneuver area. It can then support maneuver training with tactical smoke and illumination. Approximately 10 surveyed firing points are needed on each division post to calibrate aiming circles.

Natural terrain and vegetation should remain undisturbed to the maximum extent possible.

Based on safety, forward observation areas are located in front of firing areas and slightly to the flank of primary mortar to target lanes.

A common impact area is used for all types of mortars. It is at least 2,000 meters wide and 6,000 meters deep. Firing at maximum and minimum range is obtained by using different firing points.

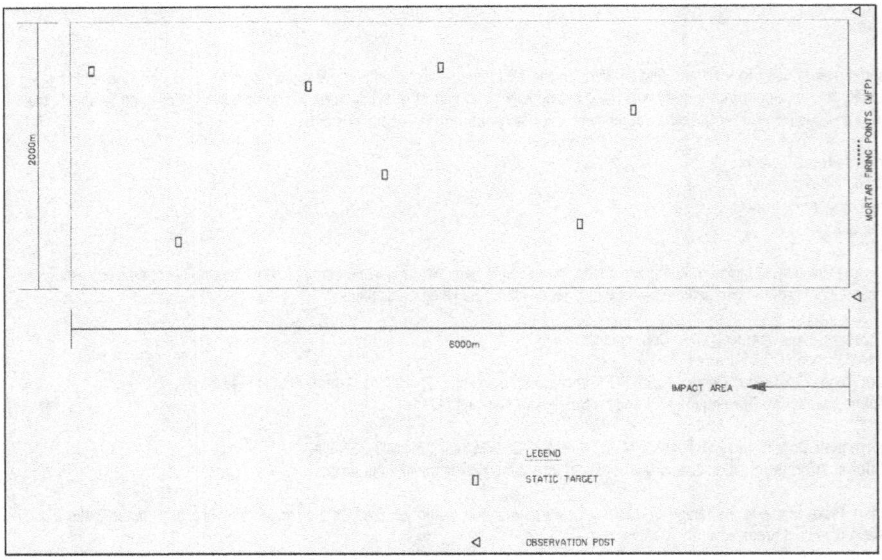

## Figure D-13. FCC 17856 FIELD ARTILLERY INDIRECT RANGE

This range is used to train field artillery crews on the skills necessary to apply fire mission data, engage, and hit stationary targets in a tactical array.

Primary features include:
Surveyed firing point(s)
Dedicated impact area(s) with hard targets

No automation is required for this facility.

Associated Range Operations and Control facilities:
No standard facilities are associated with this range.

**Requirement Document:** FM 6-30, FM 6-40, FM 6-50,TC 25-1, FM 3-09.70
**Additional Information:** Ideally, this impact area is adjacent to direct fire ranges.
The alternate firing line should be located to fit the specific site. Natural terrain and vegetation should remain undisturbed to the maximum extent possible.
The battery firing position depends on the size of the weapon, formation and terrain at the firing point. Depending on the types of artillery planned, the following data should be uses.

| Type | Dimensions (meters) | |
|---|---|---|
| M117/M102 (105mm Battery | Position | 500 wide |
| | Range | 1500 deep |
| | | 7500 wide |
| M109/M198 (155mm) Battery and M110 (203mm) Battery | Position | 1000 wide |
| | Range | 25,000 wide |
| | | 12,500 deep |
| MLRS Launcher (one weapon live fire) | Position | 10 wide |
| | Range | 15,000 wide |
| | | 30,000 deep |
| | Rear of launch blast | 800 deep |
| | | 400 wide |

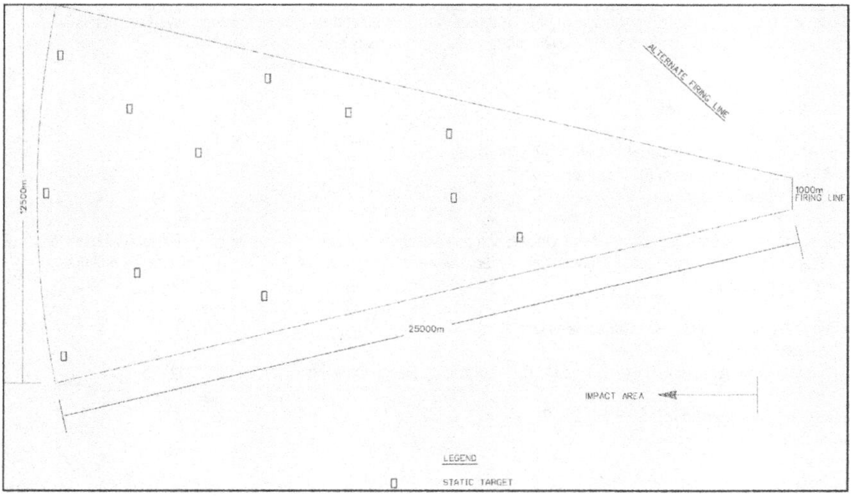

**Figure D-14. FCC 17858 SCOUT/RECONNAISSANCE GUNNERY COMPLEX (SCOUT/RECCE)**

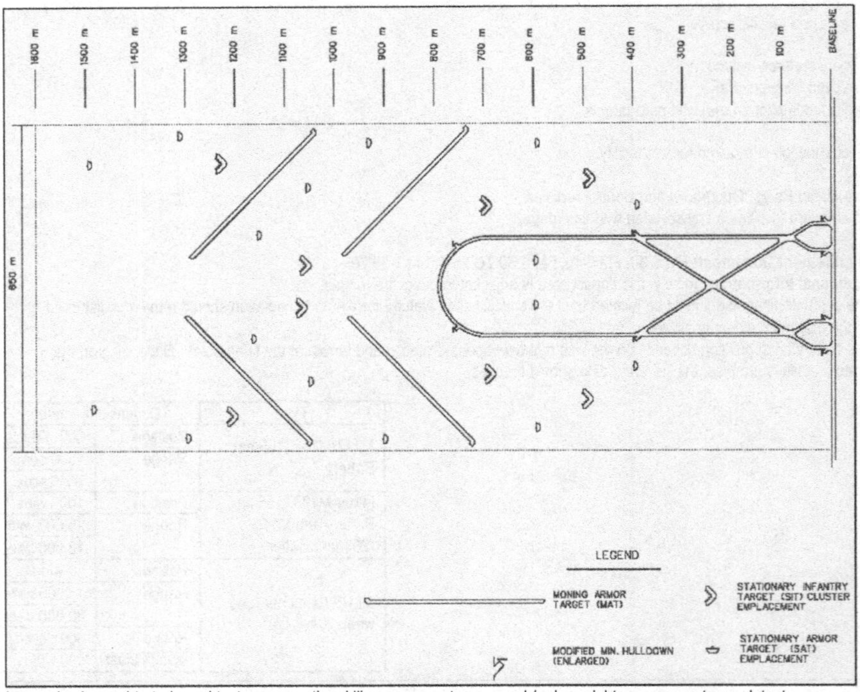


This complex is used to train and test crews on the skills necessary to zero and /or boresight weapon systems, detect, identify, engage and defeat stationary and moving infantry and armor targets in a tactical array.

Primary features include:

- 16 Stationary armor targets
- 4 Moving armor targets
- 56 Stationary infantry targets (8 clusters @ 7 SITs each)
- 1 Lane (2 course roads) with crossover capability
- 8 Battle positions per lane

All targets are fully automated, utilizing event-specific, computer-driven target scenarios and scoring. Targets will receive and transmit digital data from the range operations center. This captured data is compiled and available to the unit during the after action review.

Associated Range Operations and Control facilities:

Standard Armor ROCA Facilities

Except Range Operations Center, Large (17124) replaced by Range Operations Center, Small (17123)

**Requirement Document:** FM 3-20.8, FM 23-27

**Additional Information:** None

**Figure D-15. FCC 17859 DIGITAL MULTIPURPOSE TRAINING RANGE (DMPTR)**

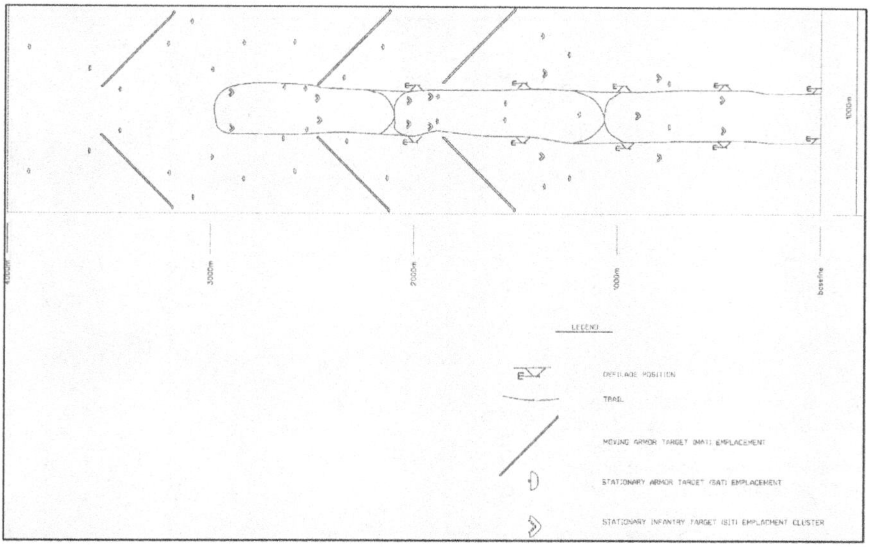


This complex is used to train and test crews and dismounted infantry squads on the skills necessary to detect, identify, engage and defeat stationary infantry and stationary/moving armor targets in a tactical array. In addition to live-fire, this complex can also be used for training with sub-caliber and/or laser training devices. The complex is specifically designed to satisfy the training and qualification requirements for the crews and sections of armor, infantry and aviation units. This complex also supports dismounted infantry squad tactical live-fire operations either independently of, or simultaneously with, supporting vehicles.

Primary features include:

- 35 Stationary armor targets
- 6 Moving armor targets
- 105 Stationary infantry targets (15 clusters @ 7 SITs each)
- 1 Lane (2 course roads) with midpoint crossover capability
- 10 Battle positions per lane

All targets are fully automated, utilizing event-specific, computer-driven target scenarios and scoring. Targets will receive and transmit digital data from the range operations center. The captured data is then compiled and available to the unit during the after action review.

Associated Range Operations and Control facilities:
Standard Armor ROCA Facilities

**Requirement Document:** FM 3-20.12, FM 23-1, FM 3-20.8, FM 3-04.140
**Additional Information:** This complex uses thermal targets, night illumination devices, and hostile-fire, target-kill, and visual flash simulators.
Location of the bore sight must be coordinated with the trainer. Gunnery tasks requiring the usage of dud producing ammunition cannot be fired on the complex. Provisions for these tasks must be made in impact areas adjacent to the complex.
Hard targets may be positioned in the impact area to facilitate RF/IR signatures for aviation gunnery.

## Figure D-16. FCC 17860 DIGITAL MULTIPURPOSE RANGE COMPLEX (DMPRC)

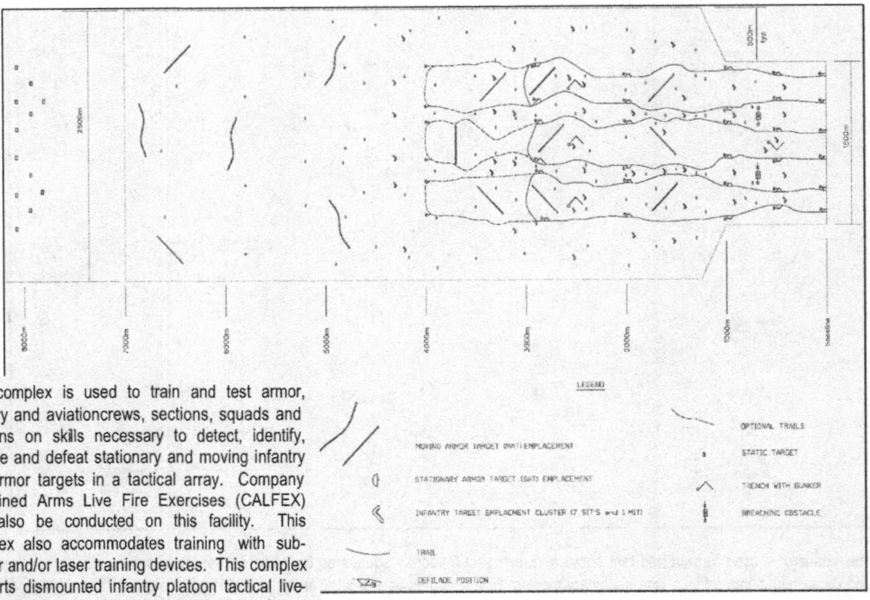


This complex is used to train and test armor, infantry and aviationcrews, sections, squads and platoons on skills necessary to detect, identify, engage and defeat stationary and moving infantry and armor targets in a tactical array. Company Combined Arms Live Fire Exercises (CALFEX) may also be conducted on this facility. This complex also accommodates training with sub-caliber and/or laser training devices. This complex supports dismounted infantry platoon tactical live-fire operations either independently of, or simultaneously with, supporting vehicles.

Primary features include:

DMPRC consists of a base area A and two optional areas B & C.

Area A - footprint is previous MPRC 1.5k x 5k with digital capability. (Meets minimum standards of armor/infantry/aviation units Tables I-XII & CALFEX)

- 69 Stationary armor targets
- 11 Moving armor targets
- 35 Infantry target emplacement clusters (245 SITs and 35 MITs total)
- 4 Trenches
- 2 Breaching obstacles
- 3 Lanes (6 course roads)
- 36 Defilade battle positions

Area B - 500m lateral extension on both sides of area A, extending to end of range (Allows greater dispersion of targets and separation of firing vehicles in order to maximize capabilities of digital platforms)

- 20 Stationary armor targets
- 10 Infantry target emplacement clusters (70 SITs and 10 MITs total)

Area C - 2k extension in depth, the width of range (Provides greater aviation capabilities and facilitates future long range direct-fire engagements from ground platforms)

- 11 Stationary armor targets
- 4 Moving armor targets
- Static targets

All targets are fully automated, utilizing event-specific, computer-driven target scenarios and scoring. Targets will receive and transmit digital data from the range operations center. The captured data is then compiled and available to the unit during the after action review.

Associated Range Operations and Control facilities:
Standard Armor ROCA Facilities
Except Small AAR (17118) replaced by Large AAR (17117);
and no General Instruction Building (17120)

**Requirement Document:** FM 3-20.12, FM 23-1, FM 3-20.8, FM 3-04.140, ARTEP 7-20 MTP, ARTEP 71-2

**Additional Information:** This complex uses thermal targets, night illumination devices, and hostile-fire, target-kill, and visual flash simulators. Location of the boresight must be coordinated with the trainer. Gunnery tasks requiring the usage of dud producing ammunition cannot be fired on the complex. Provisions for these tasks must be made in impact areas adjacent to the complex.

**Figure D-17. FCC 17865 MULTIPURPOSE TRAINING RANGE (MPTR)**

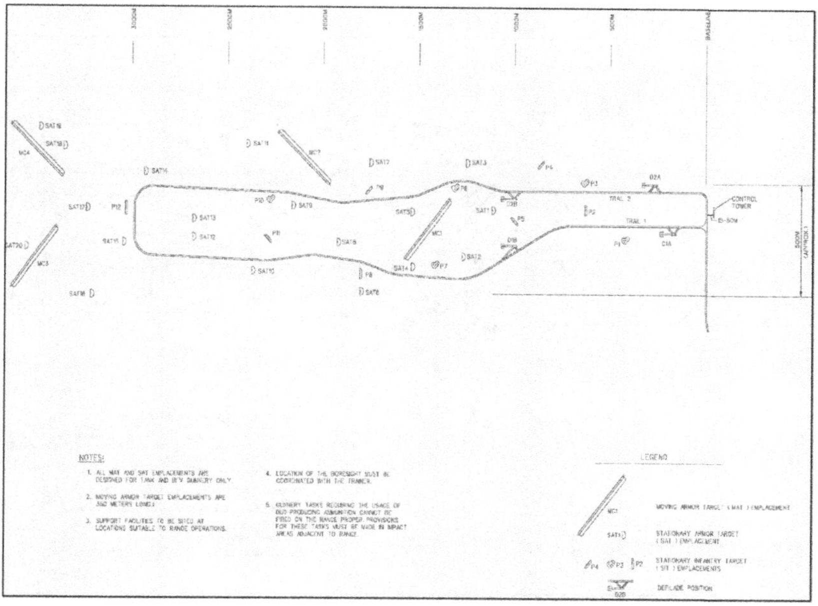

This complex is used to train and test crews and dismounted infantry squads on the skills necessary to detect, identify, and defeat stationary infantry and stationary/moving armor targets in a tactical array. In addition to live-fire, this complex can also be used for training with sub-caliber and/or laser training devices. The complex is specifically designed to satisfy the training and qualification requirements for the crews and sections of armor, infantry and aviation units. This complex also supports dismounted infantry squad tactical live-fire operations either independently of, or simultaneously with, supporting vehicles.

Primary features include:
4 Moving armor targets
10 clusters @ 7 SITs each
4 Battle positions
20 Stationary armor targets
70 Stationary infantry targets
1 Lane (2 course roads)

All targets are fully automated, utilizing event-specific, computer-driven target scenarios and scoring

Associated Range Operations and Control facilities:
Standard Armor ROCA Facilities except Range Operations Center, Large (17124) replaced by Range Operations Center, Small (17123)

**Requirement Document:** FM 3-20.12, FM 23-1, FM 3-20.8, FM 3-04.140
**Additional Information:** This complex uses thermal targets, night illumination devices, and hostile-fire, target-kill, and visual flash simulators.
Locaion of the boresight must be coordinated with the trainer. Gunnery tasks requiring the usage of dud producing ammunition cannot be fired on the complex. Provisions for these tasks must be made in impact areas adjacent to the complex.
Hard targets may be positioned in the impact area to facilitate RF/IR signatures for aviation gunnery.

**Figure D-18. FCC 17867 MULTIPURPOSE RANGE COMPLEX (MPRC)**

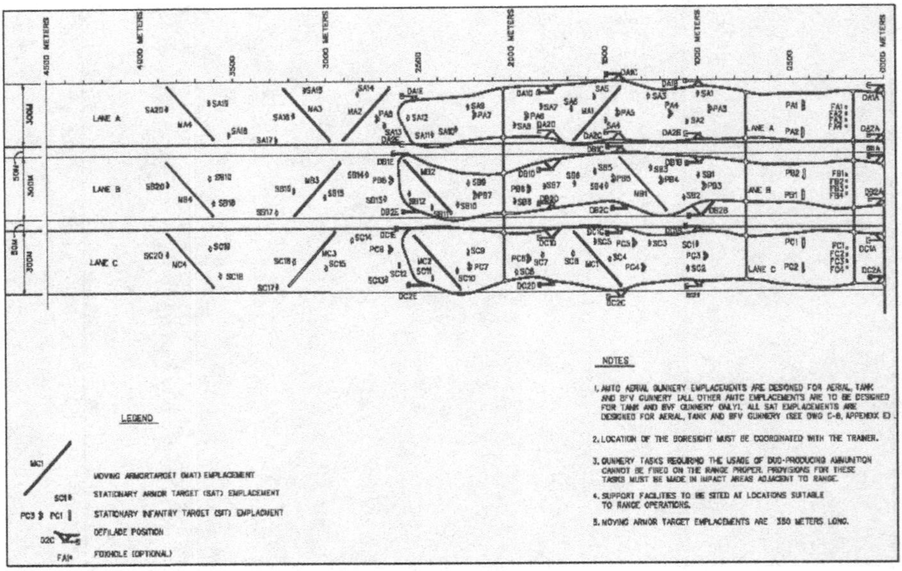


This complex is used to train and test armor, infantry and aviation crews, sections, squads and platoons on skills necessary to detect, identify, engage and defeat stationary and moving infantry and armor targets in a tactical array. This complex also accommodates training with sub-caliber and/or laser training devices.

Primary features include:

- 60 Stationary armor targets
- 12 Moving armor targets
- 154 Stationary infantry targets (22 clusters @ 7 SITs each)
- 22 Moving infantry targets
- 3 Lane (6 course roads)
- 36 Defilade battle positions

All targets are fully automated, utilizing event-specific, computer-driven target scenarios and scoring.

Associated Range Operations and Control facilities:

Standard Armor ROCA Facilities

**Requirement Document:** FM 3-20.12, FM 23-1, FM 3-20.8, FM 3-04.140, ARTEP 7-20 MTP, ARTEP 71-2
**Additional Information:** This complex uses thermal targets, night illumination devices, and hostile-fire, target-kill, and visual flash simulators. Location of the bore sight must be coordinated with the trainer. Gunnery tasks requiring the usage of dud producing ammunition cannot be fired on the complex. Provisions for these tasks must be made in impact areas adjacent to the complex.

**Figure D-19. 17870 BATTLE AREA COMPLEX (BAX)**

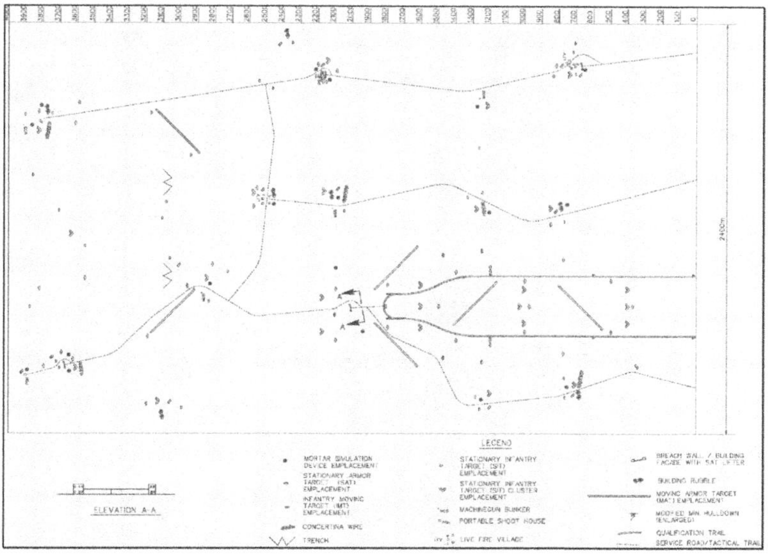


This complex provides a collective live fire home station training facility for all elements of the Stryker Brigade Combat Team (SBCT). SBCT crews and dismounted soldiers train and test their ability to detect, identify, engage and defeat stationary and moving combined arms targets in both open and urban terrain environments. This complex also supports tactical live fire operations independently of, or simultaneously with, supporting vehicles in free maneuver. This complex accommodates training with sub-caliber and/or Training devices.

Primary features include:

- 43 Stationary armor targets
- 6 Moving armor targets
- 35 Stationary infantry targets (SIT)
- 25 SIT clusters @ 7 emplacements each
- 14 Moving infantry targets
- 4 Machine gun bunkers w/sound effects simulator
- 2 Breach wall/Building facades
- 2 Live fire villages (1 with 7 buildings and 1 with 5 buildings)
- 2 Portable shoothouses
- 2 Trench lines
- 2 Lanes (2 course roads in 1 lane, no course roads in the free maneuver lane)
- 8 Hasty battle positions
- 18 Mortar simulations devices (RGSS)
- 3 Landing zones (locations not shown)

All targets are fully automated, utilizing event-specific, computer-driven target scenarios and scoring. Targets will receive and transmit digital data fom the range operations center. The captured data is then compiled and available to the unit during the after action review.

Associated Range Operations and Control facilities: Standard Armor ROCA Facilities
Except Small AAR (17118) replaced by Large AAR (17117); and no General Instruction Building (17120)

**Requirement Document:** FM 3-20.13, FM 3-20.8, FM 3-21.9, FM 3-21.11
**Additional Information:** This range uses thermal targets, night illumination devices, and hostile-fire, target-kill, and visual flash simulators.
Location of the bore sight must be coordinated with the trainer. Gunnery tasks requiring the usage of dud producing ammunition cannot be fired on the complex. Provisions for these tasks must be made in impact areas adjacent to complex.

**Figure D-20. FCC 17872 AIR DEFENSE FIRING RANGE**

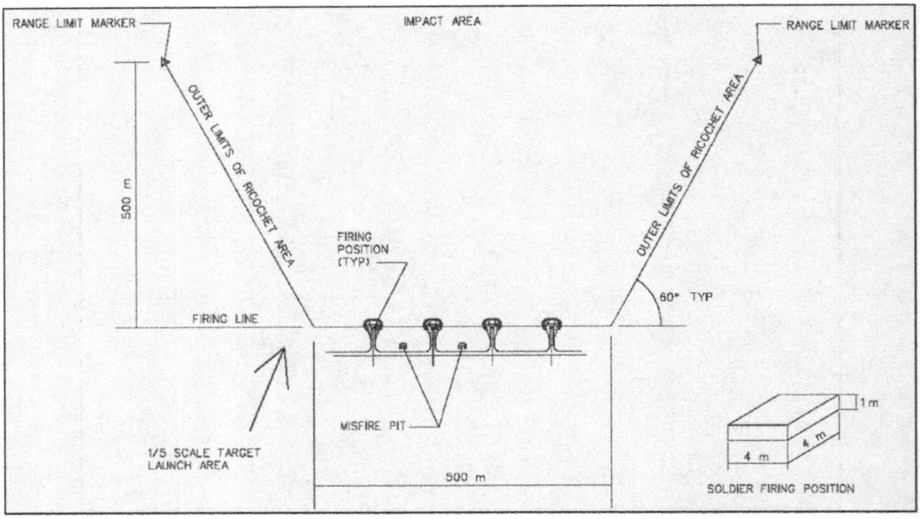


This complex is used to train and test Air Defense Artillery crews on the skills necessary to employ ground to air anti-aircraft missiles against ballistic aerial target systems (BATS) and drones. This complex supports shoulder-launched as well as vehicle-launched missiles.

Primary features include:
4 Shoulder/Vehicle launch firing positions

No automation is required for this facility.

Associated Range Operations and Control facilities:
Range Operations Center, Small (17123)
Bleacher Enclosure (75061)
Operations/Storage Building (17120)
Latrine (73075)

**Requirement Document:** FM 3-01.11, FM 44-43, FM 44-44, FM 44-18-1
**Additional Information:** For ground firing, large red arrows covered with highly visible paint are placed on right and left outer limits of the ricochet area. Red flags are placed on the firing line at start-fire and cease-fire points.

STINGER missiles have a 15 degree con-shaped danger zone centered directly to the rear of the launcher. To avoid hazards, the ground directly to the rear of the gunner is cleared of all obstacles and the launcher breech is placed one meter off the ground.

**Figure D-21. FCC 17878 URBAN ASSAULT COURSE**

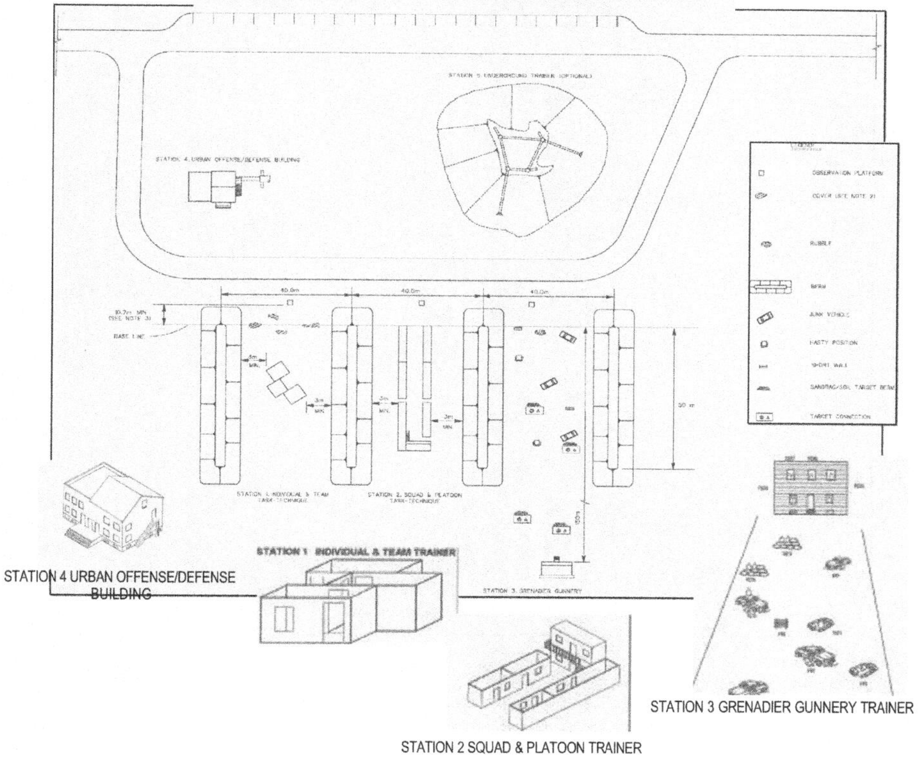


This facility is used to train individual soldiers, squads, and platoons on tasks necessary to operate within a built-up/urban area. The facility contains 5 stations:

Station 1 - Indvidual & Team Trainer
Station 2 - Squad & Platoon Trainer
Station 3 - Grenadier Gunnery Trainer
Station 4 - Urban Offense/Defense Trainer
Station 5 - Underground Trainer

Targetry Requirements:
Station 1 - 6 interior precision human urban targets (HUT)
Station 2 - 10 interior precision HUT
Station 3 –13 total targets: 6 non-precision exterior targets (SITs) and 7 non-precision fascade targets
Station 4 - 10 interior precision targets
Station 5 - no instrumentation required

All targets are fully automated and the event specific target scenario is computer driven and scored from the range operations center. Targetry in Stations 1, 2 and 4 areprecision human urban targets that have reconfigurable plug and play capability.

Associated Range Operations and Control facilities: Operations/Storage Building (17122) Latrine (73075)

**Requirement Document:** FM 3-0, FM 3-06, FM 3-06.11, ARTEP 7-8-MTP,7-3/4-1110, TC 90-1
**Additional Information:** Station 3 - 40mm HE grenades will not be used.
Station 4 - No live fire training.
Station 5 - Smoke will not be employed inside the underground trainer. Also, pyrotechnics are not authorized for station 5.

## Figure D-22. FCC 17879 LIVE FIRE EXERCISE SHOOTHOUSE

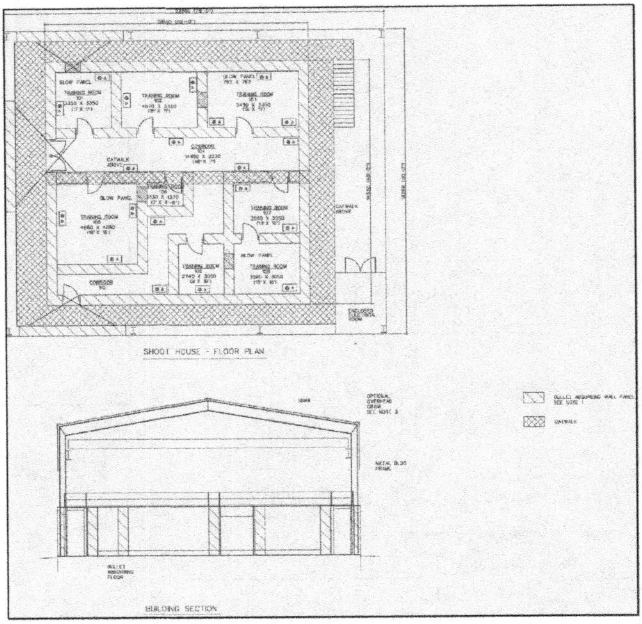


The Shoothouse provides the leader with a facility to train and evaluate the unit during a live fire exercise. Units are trained and evaluated on their ability to move tactically (enter and clear a room; enter and clear a building), engage targets, conduct breaches and practice target discrimination.

Primary features include:
Facility divided into 8 separate rooms & 2 corridors for a minimal net training capbiliby of 1400 square feet
- 20 Target/camera outlets
- 13 Precision human urban targets (HUT)

All targets are fully automated and the event specific target scenario is computer driven and scored from the range operations center. The range operating system is fully capable of providing immediate performance feedback to the using participants. All targetry are life-like precision targets that have reconfigurable plug and play capability.

Associated Range Operations and Control facilities:
Operations/Storage Building (17122)
Latrine (73075)
Small AAR Building (17118)

**Requirement Document:** TC 90-1
**Additional Information:** A roof does not reduce 360 degree SDZ, but is there to reduce light, weather protection, enhance realism and provide superstructure for an overhead crane

## Figure D-23. FCC 17880 LIVE FIRE EXERCISE BREACH FACILITY

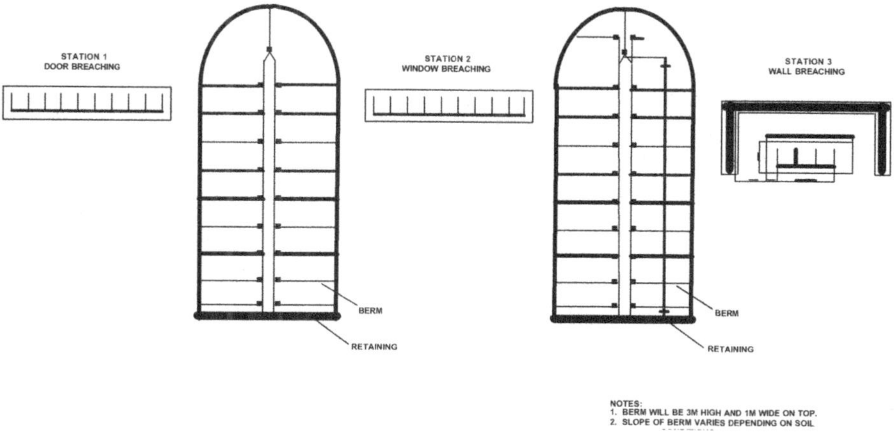


BREACH FACILTY – PLAN

The Breach Facility is used to train soldiers semi-annualy on the technical aspects of breaching techniques. It is also used to train TTP's and explosive techniques not trained on any other type of facility.

Primary features include:

- Station 1 - Door Breaching Structure
- Station 2 - Window Breaching Structure
- Station 3 - Wall Breaching Structure

No automation is required for this facility.

Associated Range Operations and Control facilities:
Latrine (73075) if not colocated with another range that has a ROCA

**Requirement Document:** TC 90-1
**Additional Information:** The trainees must retreat to a position behind the retaining wall before detonating the explosive charge.

**Figure D-24. FCC 17882 HAND GRENADE QUALIFICATION COURSE**

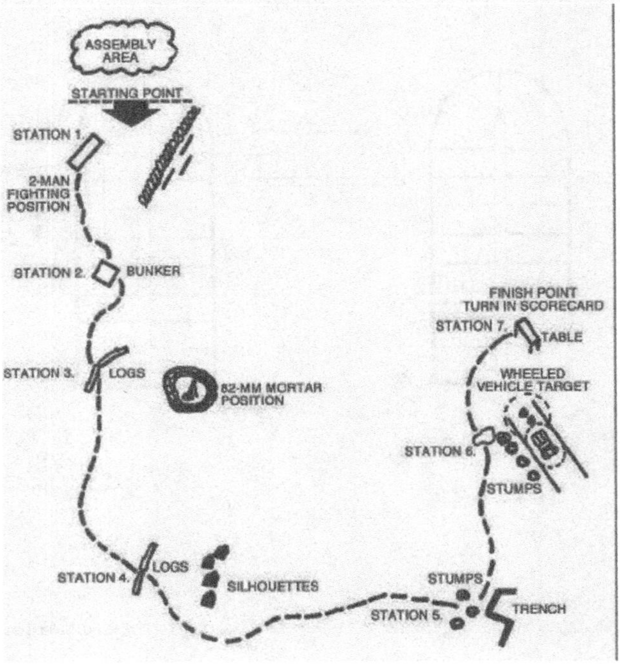


This range is used to train and test individual soldiers on the skills necessary to employ hand grenades against stationary target emplacements.

Primary features include:
7 Individual stations

No automation is required for this facility. All targets/facades are fixed at required distances.

Associated Range Operations and Control facilities:
Standard Small Arms ROCA Facilities
Except No Range Operations Center, Small (17123)

**Requirement Document:** FM 3-23.30
**Additional Information:** The qualification course allows soldiers to use fused practice hand grenades to engage targets in natural terrain under simulated combat conditions.
Station 1 - engage a group of F-type silhouettes in open from 2 man foxhole. Targets 35 meters to front beyond squad's protective wire.
Station 2 - engage a bunker with two firing portholes.
Station 3 - engage a fortified 82mm mortar position from 20 meters.
Station 4 - engage groups of targets behind cover 20 meters distance.
Station 5 - clear an entry point to a trench 25 meters distance.
Station 6 - engage troops in halted, open wheeled vehicle 25 meter distance.
Station 7 - identify hand grenades by shape, coloring, markings, and capabilities.

**Figure D-25. FCC 17883 HAND GRENADE FAMILIARIZATION RANGE**

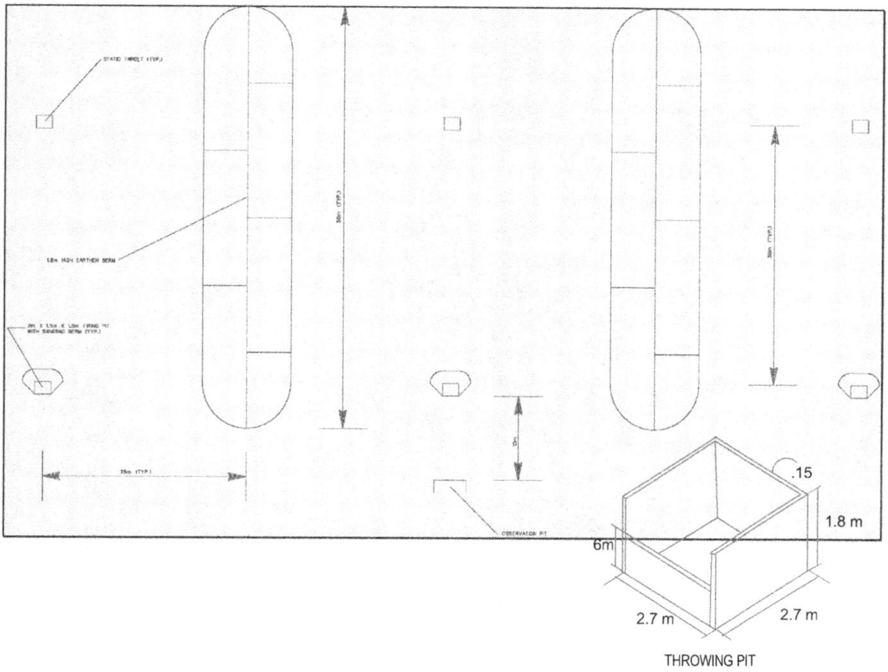


This range is used to train and test individual soldiers in the employment of live fragmentation hand grenades.

Primary features include:
- 4 Throwing bays

No automation is required for this facility. All targets are fixed at required distances

Associated Range Operations and Control facilities:
- Standard Small Arms ROCA Facilities
- Except No Range Operations Center, Small (17123)

**Requirement Document:** FM 3-23.30
**Additional Information:** Steel, concrete or wooden revetments or earthen berms 1.8m high should separate at least two of the bays from the other tow and extend out 50m. This permits practice to continue on half the range if a dud occurs. Ideally, dividers separate all of the bays.

Sand-filled drums provide an aiming point for the soldier as he throws the grenade.

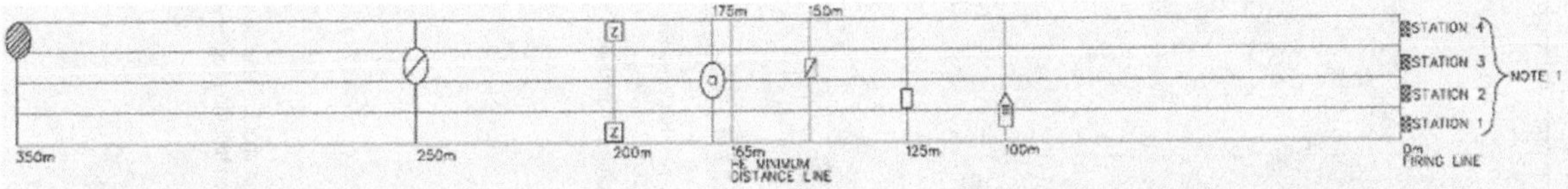


This facility is used to train and test individual solders on the skills necessary to engage and defeat stationary target emplacements with the 40mm Grenade Launcher.

Primary features include:
    4 Individual Firing Stations

No automation is required for this facility. All targets/facades are fixed at required distances.

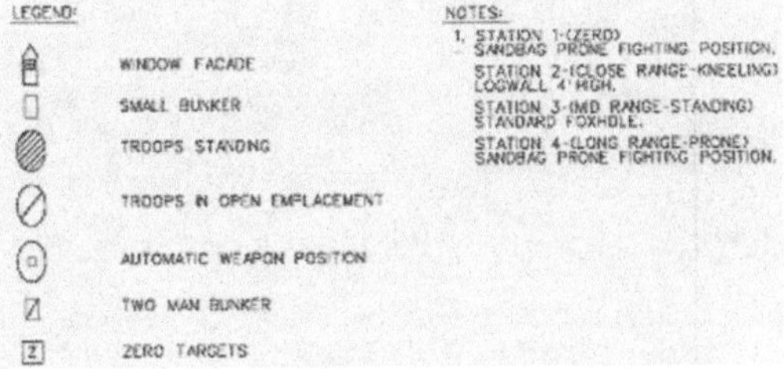


Associated Range Operations and Control facilities:
    Standard Small Arms ROCA Facilities

**Requirement Document:** FM 3-23.31, STP 21-1 SMCT
**Additional Information:** Zero targets are at least 2 meters high and 2 meters wide.
Zero target is clearly marked with a large contrasting "Z".
Targets are made of long-lasting, durable material that can withstand constant use with little maintenance. Salvaged oil drums filled with sand make excellent semi-permanent target material for this range.

**Figure D-26. FCC 17884 GRENADE LAUNCHER RANGE**

**Figure D-27. FCC 17885 LIGHT DEMOLITION RANGE**

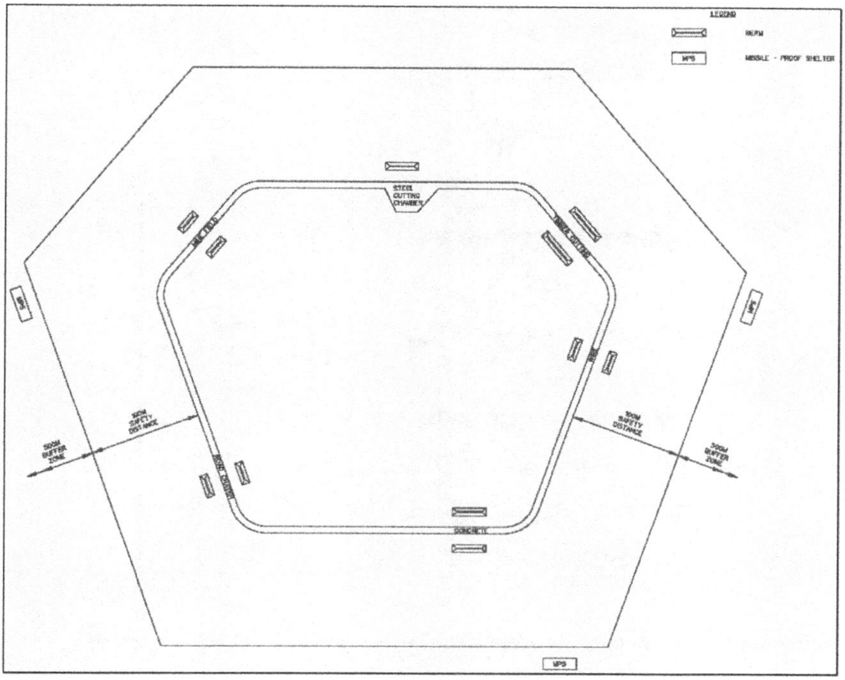


This range is used to train individual solders on the employment techniques of explosives and demolitions.

Primary features include:
- Safety berms (on each side of demolition points)
- 3 Missile-proof shelters
- Class V issue storage bunker

No automation is required for this facility.

Associated Range Operations and Control facilities:
- No ROCA facilities required.

**Requirement Document:** FM 5-250, FM 5-34, FM 5-102, FM 20-32
**Additional Information:** Wire Obstacle width - 7m; length - 20m. Use one Bangalore kit to breach the obstacle.
Minefield width - 10m; length - 20m. Mines may be placed on the surface or buried.
Steel-cutting Chamber has a ledge to mount steel or concrete beams. Steel and concrete beams vary in width and length.
Timber-cutting Site width - 10m; length - 35m. There are 8 concrete base supports (4 on each side of road) for placement of logs. Poles should be no larger than 36cm in dial nor taller than 10m. Charges should be placed to prevent damage to the concrete base.
Concrete Obstacle is constructed with concrete cubes or tetrahedrons. It is 10X30 m.
The road crater site must be refilled after each use.

**Figure D-28. FCC 17891 INFILTRATION COURSE**

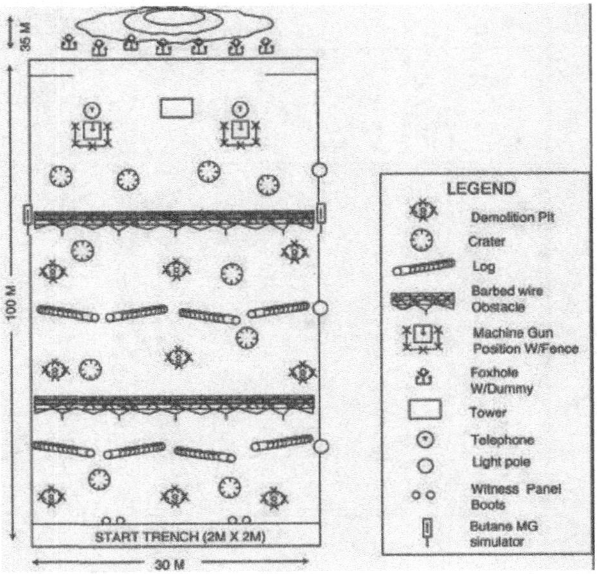


This range is used to train individual soldiers on combat movement techniques in a realistic live-fire environment.

Primary features include:
- 50 x 100 meter Movement Area
- 2 Fixed Machine Gun Positions (w/Safety Measures installed)
- 2 Barbed Wire Obstacles
- 9 Demolition Pits
- 2 Log Obstacles

No automation is required for this facility.

Associated Range Operations and Control facilities:
- Operations/Storage Building (17122)
- Bleacher Enclosure (75061)
- Latrine (73075)

**Requirement Document:** FM 7-93, FM 7-92
**Additional Information:** Construct surface danger fans for each machine gun; establish firing procedures and overhead clearance of personnel in accordance with AR/DA Pam 385-63.

When the range is opened, test fire each machine gun. Conduct test firings before each use of the range to determine dispersion pattern. Use machine gun traverse and elevating stops to prevent firing beyond the right and left limits of fire or below minimum elevations.

## Figure D-29. FCC 17892 FIRE AND MOVEMENT RANGE



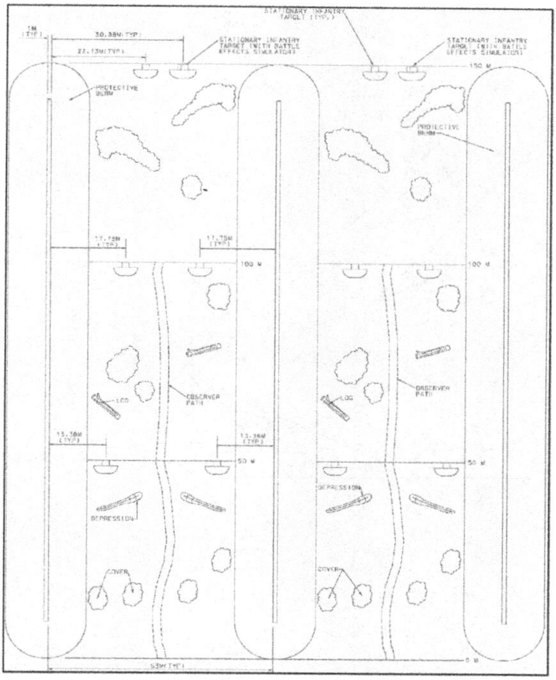


This range is used to train individual soldiers and buddy/teams on basic fire and movement techniques against stationary infantry targets. Soldiers show their ability to select covered and concealed positions, move while under fire, apply principles of teamwork, and use suppressive fire on enemy soldier targets.

Primary features include:
- 4 Lanes
- 6 Stationary infantry targets per lane
- 3 meter high berms along each side of lanes

Targets are not fully automated and/or the scenarios are not computer driven or scored.

Associated Range Operations and Control facilities:
- Range Operations Center, Small (17123)
- Operations/Storage Building (17122)
- Bleacher Enclosure (75061)
- Ammo Breakdown Building (17129)
- Latrine (73075)

**Requirement Document:** FM 7-8, TC 7-9
**Additional Information:** All lanes should have natural features that offer cover and concealment.

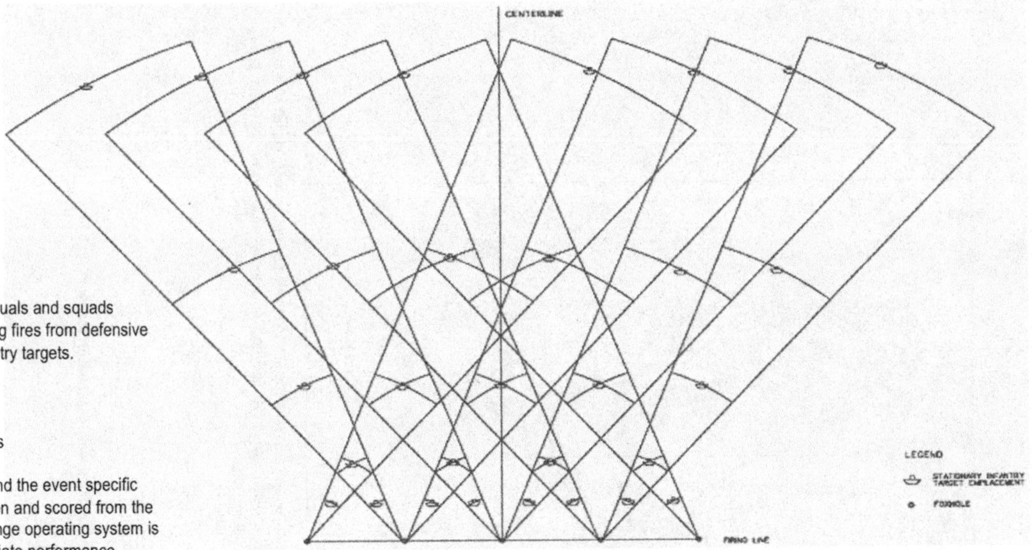


This range is used to train individuals and squads on employing mutually supporting fires from defensive positions against stationary infantry targets.

Primary features include:
- 5 Two-man foxholes
- 31 Stationary infantry targets

All targets are fully automated and the event specific target scenario is computer driven and scored from the range operations center. The range operating system is fully capable of providing immediate performance feedback to the using participants.

Associated Range Operations and Control facilities:
Standard Small Arms ROCA Facilities

**Requirement Document:** FM 7-8, TC 7-9
**Additional Information:** Each defensive position will have a right and a left sector of fire.

**Figure D-30. FCC 17893 SQUAD DEFENSE RANGE**

**Figure D-31. FCC 17895 INFANTRY SQUAD BATTLE COURSE**

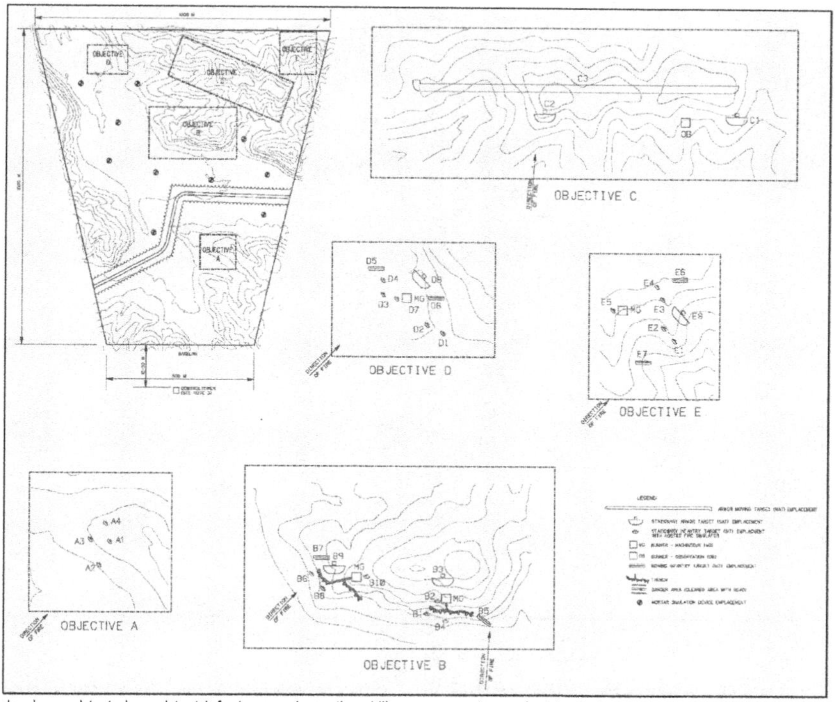


This complex is used to train and test infantry squads on the skills necessary to conduct tactical movement techniques, detect, identify, engage and defeat stationary and moving infantry and armor targets in a tactical array

Primary features include:
- 1 Moving armor targets
- 6 Moving infantry targets
- 2 Trench obstacles
- 6 Stationary armor targets
- 20 Stationary infantry targets
- 5 Machine gun/Observation bunkers (w/sound effects simulator)

All targets are fully automated and the event specific target scenario is computer driven and scored from the range operations center. The range operating system is fully capable of providing immediate performance feedback to the using participants.

Associated Range Operations and Control facilities:
Standard Small Arms ROCA Facilities

**Requirement Document:** FM 7-8, ARTEP 7-8-MTP, ARTEP 7-8 Drill, TC 7-9
**Additional Information:** Target locations are site adapted. All must be located in areas that support desired tactics and the user's training requirements.

All trenches, bunkers, and target emplacements must simulate typical threat scenarios.
Mortar simulation device emplacements are located in areas from which unfriendly mortar fire is to be simulated.
Helicopter landing areas, designed for heavy use, should be located to support aerial insertion and extraction of the squad.

**Figure D-32. FCC 17897 INFANTRY PLATOON BATTLE COURSE**

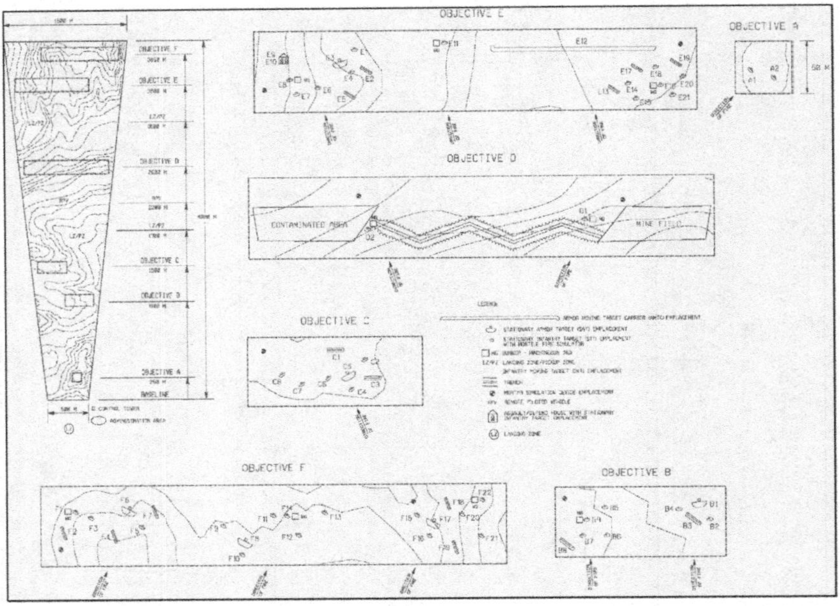


This complex is used to train and test infantry platoons, either mounted or dismounted, on the skills necessary to conduct tactical movement techniques, detect, identify, engage and defeat stationary and moving infantry and armor targets in a tactical array

Primary features include:
- 1 Moving armor targets
- 14 Moving infantry targets
- 1 Trench obstacles
- 2 Landing zones
- 6 Stationary armor targets
- 43 Stationary infantry targets
- 9 Machine gun bunkers (w/sound effects simulator)
- 1 Assault/defend house

All targets are fully automated and the event specific target scenario is computer driven and scored from the range operations center. The range operating system is fully capable of providing immediate performance feedback to the using participants.

Associated Range Operations and Control facilities:
Standard Small Arms ROCA Facilities

**Requirement Document:** FM 7-8, ARTEP 7-8 Drill, ARTEP 7-8 MTP, FM 23-1, FM 3-20.8, TC 7-9
**Additional Information:** Target locations are site adapted. All must be located in areas that support desired tactics and the user's training requirements.

All trenches, bunkers, and target emplacements must simulate typical threat scenarios.
Mortar

**Figure D-33. FCC 17901 COMBINED ARMS COLLECTIVE TRAINING FACILITY (CACTF)**

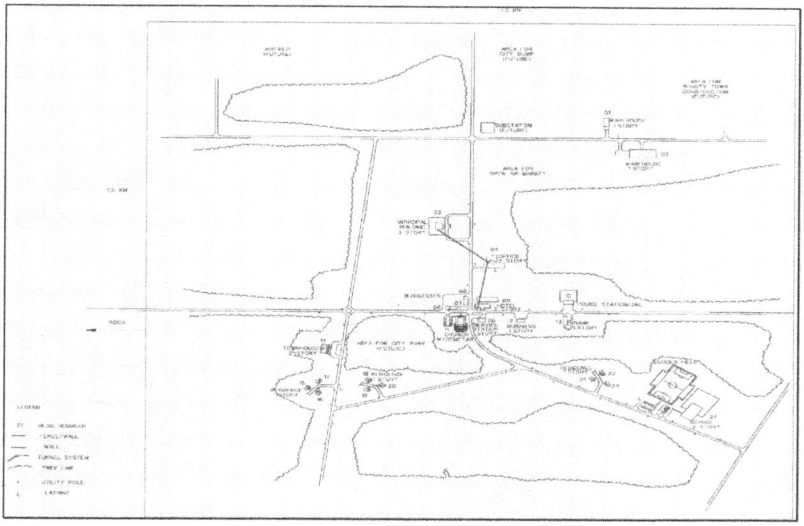

This facility is designed to conduct multi-echelon, full spectrum operations training up to battalion TF level. The CACTF will accommodate Force on Force (FOF) and Force on Targetry (FOT).

Primary features include:
1 School
1 Church/cemetery
1 Police station/Jail
1 Hotel
9 Residences
4 Businesses
1 Townhouse
1 Bank
2 Warehouses
1 Government Building
1 Office
1 Service Station

Recommended features Include:
Tunnel/Sewer System
Town Shanty
One, three-story building
Three, two-story buildings
Breachable walls
Loopholes
Mouse holes

Targets:
15 Precision/Human urban targets (HUT)
30 SITs (Recommend RF/Reconfigurable)
9 SATs (Recommend RF/Reconfigurable)

Associated Range Operations and Control facilities:
Range Operations Center, Large (17124)
Operations/Storage Building (17122)
Latrine (73075)
Covered Mess (17116)
Large AAR (17117)

All targets are fully automated, utilizing event-specific, computer-driven target scenarios and scoring. Targets will receive and transmit data from the range operations center. The captured data is then compiled and available to the unit during the after action review. All targetry are life-like precision targets that have reconfigurable plug and play capability.

Camera Coverage Requirement:
80% of Buildings with 80% room coverage; considerations should be given to hallways and stairwells.

Live-fire ballistic ammunition is prohibited on this facility. Pyrotechnics are not authorized for training in the tunnel system.

**Requirement Document:** TC 90-1, FM 3-22.9
**Additional Information:** The ROCA should not be located in the immediate vicinity of the urban training area.

**Figure D-34. AROCA STANDARD ARMOR RANGE OPERATIONS AND CONTROL AREA FACILITIES**

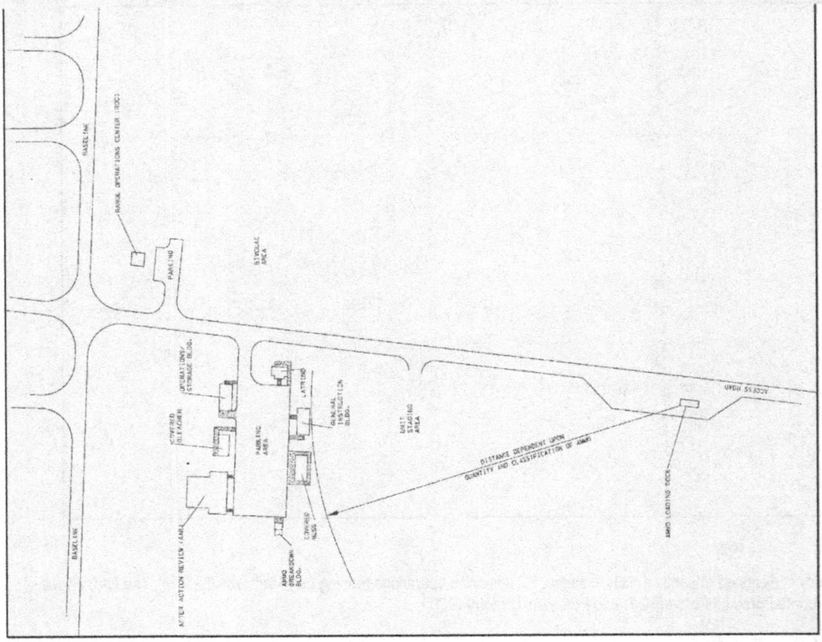


The Range Operation Center and Operations/Storage Building are used to operate and maintain the range. The Bleacher Enclosure, General Instruction and After Action Review buildings are used for pre and post event instruction. The remaining buildings are to support the training or the troops being trained.

Associated Range Operations and Control facilities:
Range Operations Center, Large (17124)
Operations/Storage Building (17122)
General Instruction Building (17120)
Latrine (73075)
Bleacher Enclosure (75061)
Covered Mess (17116)
Ammo Breakdown Building (17129)
After Action Review Building, Small (17118)
Ammunition Loading Dock (14970)
Bivouac Area
Unit Staging Area (85212)

**Additional Information:** The drawing depicts a typical layout and may be adjusted for project requirements.
Minimum set back distances of building from the baseline are mandated in AR 385-63 and AR 385-64.

**Figure D-35. SROCA STANDARD SMALL ARMS RANGE OPERATIONS AND CONTROL AREA FACILITIES**

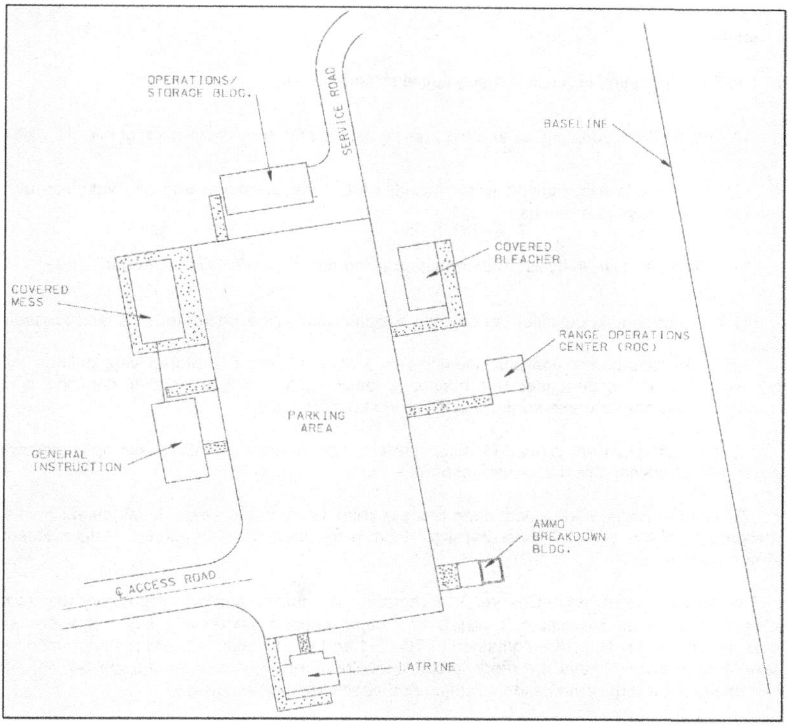


The Range Operation Center and Operations/Storage Building are used to operate and maintain the range. The Bleacher Enclosure and General Instruction Building are used for pre and post event instruction. The remaining buildings are to support the training or the troops being trained.

Associated Range Operations and Control facilities:
Range Operations Center, Small (17123)
Operations/Storage Building (17122)
General Instruction Building (17120)
Latrine (73075)
Bleacher Enclosure (75061)
Covered Mess (17116)
Ammo Breakdown Building (17129)

**Additional Information:** The drawing depicts a typical layout and may be adjusted for project requirements.
Minimum set back distances of building from the baseline are mandted in AR 385-63 and AR 385-64.

# Appendix E
# RLTP Planning Process

## E-1. Process

a. The RTLP planning process is a sequential six-step process:

(1) The "asset accounting" of existing training assets at a given installation or site.

(2) The "needs assessment" which includes METL requirements and an evaluation of range conditions to perform training missions.

(3) A "comparative analysis" to determine training asset shortfall(s) or adequacy.

(4) A "requirements definition" to develop specific capabilities, characteristics, and functionality.

(5) An "alternative analysis" identifies all viable or potential alternatives; defines mission support, environmental consequences and economic feasibility for each alternative identified; and site planning to determine any land use conflicts or environmental impacts.

(6) The "implementation plan" identifies preferred alternatives prioritized and an implementation timeline to establish critical milestones and resources.

(7) The requirement identified in the process culminating in the range development plan serves as the description of the mission requirement for land in the installation Integrated Natural Resources Management Plan (INRMP).

b. Planners (as identified in Chapter 1, paragraph 1-4) must comply with regulatory requirements, environmental matters, and examine a variety of factors when considering a site. Specific factors, procedures, and related policies are contained in TC 25-1 and this circular. Should the alternative require the development of a range, see the range project development process in paragraph 3-4, AR 210-21. Additional guidance on requirements identification and needs assessment follow.

## E-2. Requirements Identification

a. Doctrine, force structure, weapon systems, and mission tasks are evaluated to determine the specific requirements necessary to train soldiers and units to meet operational readiness and Army training goals. Field Manuals provide the foundation for development of the unit METL. The following documents will assist in identifying requirements:

(1) Field Manuals 7-0 and appropriate type FMs assist in the development of METL that focus on those training tasks which are essential to accomplishing an organization's wartime mission.

(2) Unit ARTEPs and CATS provide unit-training requirements to be executed in maneuver areas and on live-fire ranges.

(3) Weapon system training strategies and standards in respective field manuals.

(4) Department of the Army Pamphlet 350-38, outlines recommended weapons training programs, qualification standards, training ammunition, operating tempo and suggested training strategies.

(5) Support Facility Annex (SFA) to the Integrated Logistics Support Plan published by HQUSACE identifies facility implications associated with material fielding, including training considerations and is continuously updated as the weapon system matures. These SFAs are available to

installations and MACOM via the Programming, Administration, and Execution (PAX) system under the Facility Planning System utility.

(6) Training Circular 25-1, this circular, and AR/DA Pam 385-63 provide guidance on site planning, range layouts, safety standards, SDZs, standard range facilities, training land, instrumentation, and specific weapon system requirements.

**E-3. Needs Assessment**

a. Planners must evaluate existing capabilities to meet standard training readiness requirements, alternative methods to maximize training effectiveness, and resources (for example, ammunition, fuel, training land and range facilities, manpower, life cycle O&M) to meet operational readiness. Inclusive is an assessment of range and training land effectiveness, conditions, utilization factors, and environmental influences to perform training missions. During the STRAC process, training strategies are developed incorporating combinations of devices, simulators and live fire requirements to satisfy training proficiency requirements in a holistic manner. Potential environmental restrictions related to range standardization or modernization, and training land requirements must be identified during the needs assessment and may mandate a tailored training strategy to accommodate existing training land restrictions. The following documents will assist in assessing potential range construction and training land needs:

(1) Army Regulations 200-1, 200-3, 200-4, and 210-20, and TC 25-1 outline planning considerations to be addressed during this assessment.

(2) Army Regulations 200-2, 415-15, 415-20, and the Master Planning Instructions outline technical procedures for planning, programming, estimating construction costs and milestones required to either construct or acquire facilities.

(3) United States Army Corps of Engineers (CEHNC 1110-1xx series) design manuals depict standardized facilities.

(4) AR 385-10 and DA PAM 385-63 outline range safety considerations and SDZ requirements.

b. Potential impacts of noise from a new or modified live-fire range on the surrounding civilian community can inhibit full utilization during the conduct of training. Environmental considerations as defined in AR 200-1 and applicable to Installation Compatible Use Zones will be evaluated when modifying existing ranges or siting new ranges.

c. The ITAM Training Requirements Integration (TRI) process is employed to identify optimum siting of training ranges and facilities and optimum modifications to training areas.

# Glossary

## Section I
## Abbreviations

| | |
|---|---|
| A | area target |
| AAR | after-action review |
| AATLFR | antiarmor tracking and live fire range |
| ADF | air defense firing |
| AE | architectural engineer |
| AFF | automated field-fire |
| AGES/AD | Air-ground engagement system/air defense |
| AGR | aerial gunnery range |
| AH | attack helicopter |
| AL | Alabama |
| AMC | US Army Materiel Command |
| AMRD | Army Modernization Reference Data |
| AP | armor-piercing |
| approx | approximately |
| AR | Army regulation |
| ARF | automated record-fire |
| ARTEP | Army Training and Evaluation Program |
| ATS | Army target system |
| ATTN | attention |
| BATS | ballistic aerial target system |
| BCTP | Battle Command Training Program |
| BES | Battlefield Effects Simulator |
| BFV | Bradley Fighting Vehicle |
| Btry | Battery |
| C | civilian; troops |
| CALFEX | combined arms live-fire exercise |
| cal | caliber |
| CATS | Combined Arms Training Strategy |
| CCTT | close combat tactical trainer |
| CD-ROM | compact disk – read only memory |
| CEHNC | Corps of Engineers, Huntsville Engineering and Support Center |
| CEHNDM | Corps of Engineers, Huntsville Division, manual |
| CEV | combat engineer vehicle |
| CFV | Cavalry Fighting Vehicle |
| CH | Cargo helicopter |
| cm | centimeter |
| CMTC | Combat Maneuver Training Center |
| Co | Company |
| COE | Corps of Engineers |
| cont | continued |
| CONUS | continental United States |
| CPQC | combat pistol qualification course |
| CPT | captain |
| CS | combat support |
| CSS | combat service support |
| CTC | combat training center |
| DA | Department of the Army |
| DCSOPS | Deputy Chief of Staff for Operations and Plans |

| | |
|---|---|
| def | defensive |
| DMPRC | digital multipurpose range complex |
| DPTM | directorate of plans, training, and mobilization |
| DPW | directorate of public works |
| DRM | directorate of resource management |
| DSN | Defense Switched Network |
| DTLOMS | doctrine, training, leader development, organization, materiel, and soldiers |
| | |
| EA | environmental assessment |
| EIS | environmental impact statement |
| EOD | explosive ordnance disposal |
| EPLRS | Enhanced Position Location Reporting System |
| | |
| FBCB2 | Force XXI Battle Command Brigade and Below |
| FEE | functional estimating equations |
| FM | field manual; frequency modulated |
| FP | firing position |
| FTX | field training exercise |
| | |
| G3 | Assistant Chief of Staff, G3 (Operations and Plans) |
| GIS | geographic information system |
| GLLD | Ground laser locator designator |
| GS | General Schedule |
| GTA | Grafenwoehr Training Area |
| GTL | gun-target line |
| | |
| HE | high explosive |
| HG | hand grenade |
| HMMWV | high-mobility multi-purpose wheeled vehicle |
| HQDA | Headquarters, Department of the Army |
| HTA | Hohenfels Training Area |
| | |
| IBCT | Interim Brigade Combat Team |
| IHFS | infantry hostile-fire simulator |
| IN | inch |
| INRMP | Integrated Natural Resources Management Plan |
| IPBC | infantry platoon battle course |
| IR | infrared |
| ISBC | infantry squad battle course |
| ITAM | Integrated Training Area Management |
| ITAS | Improved Target Acquisition System |
| ITV | Improved TOW Vehicle |
| | |
| JAAT | joint air-attack team |
| JRTC | Joint Readiness Training Center |
| | |
| kmh | kilometers per hour |
| | |
| LAV | light armored vehicle |
| LAW | light anti-tank weapon |
| lb | pound |
| LD | line of departure |
| LFX | live-fire exercise |
| LOS | line-of-sight |
| LSO | laser safety officer |
| LT | lieutenant |

| | |
|---|---|
| LTA | local training area |
| LURS | land use requirements study |
| LZ | landing zone |
| | |
| M | military/meter |
| MACOM | major Army command |
| MAJ | major |
| MAT | moving armor target |
| MD | Maryland |
| METL | mission-essential task list |
| MFS | Muzzle Flash Simulator |
| MG | machine gun |
| MILES | multiple integrated laser engagement system |
| MIT | moving infantry target |
| MLRS | multiple launch rocket system |
| mm | millimeter |
| MOS | military occupational specialty |
| MOUT | military operations on urbanized terrain |
| MPB | Master Planning Board |
| MPFQC | military police firearms qualification course |
| MPH | miles per hour |
| MPMG | multi-purpose machine gun |
| MPRC | multipurpose range complex |
| MPS | missile-proof shelter |
| MPTR | multipurpose training range |
| MRA | maneuver rights area |
| MRF | modified record fire |
| MSD | MILES Shoot back Device |
| MS-3 | Manpower Staffing Standards System |
| MTA | major training area |
| MTP | mission training plan |
| MULE | modular universal laser equipment |
| | |
| NA | not applicable |
| NBC | nuclear, biological, chemical |
| NCO | noncommissioned officer |
| NCOIC | noncommissioned officer-in-charge |
| NdYAG | neodymium doped yttrium aluminum garnet |
| NGB | National Guard Bureau |
| no | number |
| NSN | national stock number |
| NTC | National Training Center |
| | |
| Obi | objective |
| OC | observer-controller |
| OE-CX | Ordnance and Explosives Center of Expertise |
| off | offensive |
| OIC | officer in charge |
| O&M | operation and maintenance |
| OPFOR | opposing force |
| | |
| P | point target |
| Pam | pamphlet |
| PEO | program executive officer |
| PGTS | precision gunnery target system |
| Plt | platoon |

| | |
|---|---|
| PM | project manager |
| POTA | Patriot omni-directional training, aerial |
| PPBES | planning, programming, budgeting, and execution system |
| PZ | pickup zone |
| QTR | qualification training range |
| RADS | range analysis and design system |
| RC | Reserve Component |
| RCMAT | radio-controlled miniature aerial target |
| RCS | radar cross section/Range Control System |
| RDP | range development plan |
| Recon | reconnaissance |
| RETS | remoted target system |
| RFMSS | Range Facility Management Support System |
| RLEM | rifle launched entry munition |
| RNG-MCX | Range-Mandatory Center of Expertise |
| RPV | remotely piloted vehicle |
| RSO | range safety officer |
| RTLP | Range and Training Land Program |
| RTLP-MCX | Range and Training Land Program – Mandatory Center of Expertise |
| S | Army wide standard |
| SAT | stationary armor target |
| SAW | squad automatic weapon |
| SDZ | surface danger zone |
| SES | Sound Effects Simulator |
| SESAMS | special effects small arms marking system |
| SFA | support facility annex |
| SGR | scaled gunnery range |
| SIT | stationary infantry target |
| SMR | scaled mortar range |
| SOP | standing operating procedure |
| SRTA | short range training ammunition |
| STRAC | Standards in Training Commission |
| STX | situational training exercise |
| T | tank |
| T&EO | training and evaluation outline |
| TADS | target acquisition detection system |
| TADSS | training aids, devices, simulators, and simulations |
| TBD | to be determined |
| TC | training circular; throughput capacity |
| TCPC | tank crew proficiency course |
| TES | tactical engagement simulation |
| TGT | target |
| THP | take-home package |
| TOW | tube-launched, optically-tracked, wire-guided |
| TP | target practice |
| TR | throughput requirement |
| TRADOC | US Army Training and Doctrine Command |
| TRC | training readiness condition |
| TRI | training requirements integration |
| TSM | TRADOC system manager |
| TTM | tank target mechanism |
| TTP | tactics, techniques, and procedures |

| | |
|---|---|
| TWGSS/PGS | tank weapons gunnery simulation system/precision gunnery system |
| UCOFT | Unit Conduct of Fire Trainer |
| UH | utility helicopter |
| US | United States |
| USACE | US Army Corps of Engineers |
| USAESCH | US Army Engineering Support Center, Huntsville |
| USAEDH | US Army Engineer Division, Huntsville |
| USAF | US Air Force |
| USAREUR | US Army, Europe |
| USATSC | US Army Training Support Center |
| VA | Virginia |
| VIP | very important person |
| WB | wage board |
| WG | wage grade |

## Section II
## Terms

**Active Army**
The active Army consists of (1) members of the Regular Army on active duty; (2) members of the Army National Guard of the United States and Army Reserves on active duty (other than for training); (3) members of the Army National Guard in the service of the United States pursuant to a call; and (4) persons appointed, enlisted, or inducted into the Army without component.

**After-action review (AAR)**
A professional discussion that focuses on the objectives of ongoing or completed training. It reviews a training activity to permit participants to discover for themselves what happened and why.

**Army National Guard**
The Army portion of the organized militia of the states, Commonwealth of Puerto Rico, and the District of Columbia whose units and members are federally recognized.

**Army National Guard of the United States**
A reserve component of the Army, all of whose members are members of the Army National Guard.

**Category code**
A numerical code used to classify and categorize Army real property. The Army generally uses a five-digit code to plan, program, budget, design, construct, inventory, and maintain its facilities.

**Collective training**
Training either in institutions or units that prepares cohesive teams and units to accomplish their missions on the battlefield and in operations other than war.

**Combined arms live-fire exercise (CALFEX)**
A combat exercise in which a combined-arms team, in combat formation, conducts a coordinated combat firing and maneuver practice to assault, seize, and defend appropriate objectives. The CALFEX may include tactical air support.

**Cook off**
A functioning of any of the ammunition explosive components because of high temperatures within the weapon.

**Crew-served weapon**
A weapon system for which a crew of at least two members is required.

**Force XXI Battle Command Brigade and Below**
A digital command-and-control system that provides battle command and situational awareness information from brigade down to the soldier and platform level.

**Hang fire**
A delay in functioning of a weapon.

**Impact area**
The land area and associated airspace within a training complex intended to capture or contain ammunition, munitions, or explosives to include resulting debris, fragments, and components from weapon system employment.

**Installation**
An aggregation of contiguous or near-contiguous, common mission-supporting real property holdings under the jurisdiction of the Department of Defense or a state, the District of Columbia, territory, commonwealth, or possession, controlled by and at which a Department of Defense unit or activity (active or Reserve Component) is permanently assigned. A fixed location together with its land, buildings, structures, utilities, and improvements.

**Misfire**
A failure to fire, not necessarily hazardous. Since it cannot readily be distinguished from a delay in functioning (hang fire), it must be handled as worst case in accordance with procedures for the weapon system.

**Mission Training Plan (MTP)**
A guide for units on what and how to train to achieve mission proficiency. It identifies unit wartime missions and the critical tasks that support them, provides comprehensive training and evaluation outlines, and provides exercises and other management aids to assist field commanders in planning and executing unit training.

**Range**
An area that is reserved and normally equipped for practice in weapons delivery or shooting at targets.

**Reserve Components (RC)**
The US Army Reserve (USAR) and the US Army National Guard (ARNG).

**Stryker Brigade Combat Team (SBCT)**
The Army's evolving structure – called the "transformation force." The restructured Army will have improved deployment characteristics while maintaining the capability to provide humanitarian assistance and prosecute military operations in low to high intensity conflicts.

**Sub caliber ammunition**
Practice ammunition of a caliber smaller than standard for the gun on which practice is being given. Sub caliber ammunition is economical and may be fired in relatively crowded areas. It is used with special sub caliber equipment to simulate firing conditions with standard ammunition.

**Surface danger zone (SDZ)**
The area designated on the ground of a training complex (to include associated safety areas) for the vertical and lateral containment of projectiles, fragments, debris, and components resulting from the firing or detonation of weapon systems. Army Regulation and Department of the Army Pamphlet 385-63 describes and illustrates weapon-system SDZs.

**Sustainment training**

Individual and collective training conducted in the unit or resident school to ensure continued expertise on the operations, employment, and logistics support of fielded systems or equipment.

**Tactical engagement simulation training system**
An advanced collective training methodology supported by a family of TADSS used in free-play force-on-force, field training exercises. The TES training system consists of three subsystems – the simulator subsystem which includes TADSS and supporting procedures that simulate casualty-producing effects of weapons in real time; the control subsystem which includes a staff of trained observer-controllers who referee, ensure realism, record events, and report actions observed through AARs and unit take-home packages; and the management subsystem which includes activities and computers to plan, schedule, conduct, and evaluate the training of battle-focused METL tasks to standard.

**Take-home package (THP)**
A compilation of a unit's experience at a combat training center (CTC) provided to the unit after completing training at the CTC. The observer-controllers compile the THP using engagement and other data collected by the CTC instrumentation system. The THP lets the unit replay a training exercise at an LTA or MTA, focusing on areas that need improvement to achieve and sustain readiness.

**Throughput**
The number of individuals, crews, or units that use a range. The number of individuals, crews, or units required to train is the throughput requirement. The number of individuals, crews, or units that can accomplish all required iterations of training on a given range during a single year is the annual throughput capacity of the range. When comparing the annual throughput requirement and capacity, if the throughput capacity exceeds the throughput requirement of a given range, there exists excess capacity. If the throughput requirement exceeds the throughput capacity, a need exists for additional training capability.

**Training aids, devices, simulators, and simulations (TADSS)**
A term that includes training range and combat training center (CTC) instrumentation, tactical engagement simulation (TES), battle simulation, targets, training-unique ammunition, and dummy, drill, and inert munitions. TADSS are subject to the public laws and regulatory guidance governing the acquisition of materiel.

**Training land**
The term "training land" encompasses facilities (e.g., ranges, maneuver land, proficiency courses, and direct-support facilities) dedicated to preparing and sustaining personnel and units to meet mission roles and standards. Examples include facilities to support training personnel in weapon systems use and proficiency, occupational skills, and standards developed.

**References**

**Section I**
**Sources**

These are the sources quoted or paraphrased in this publication.

AR 210-20. *Master Planning for Army Installations.* 30 Jul 93.
AR 210-21. *Army Ranges and Training Land Program.* 1 May 97.
AR 350-38. *Training Device Policies and Management.* 15 October 1993.
AR 385-40. *Accident Reporting and Records.* 1 Nov 94.
AR 385-63/MCO P3570.1B. *Range Safety.* 19 May 03.
AR 415-28. *Real Property Category Codes.* 10 Oct 96.
DA Pam 350-9. *Index and Description of Army Training Devices.* 3 Sep 2002
DA Pam 350-38. *Standards in Weapon Training.* 3 Jul 97.
DA Pam 385-63. *Range Safety.* 10 Apr 03
DA Pam 415-28. *Guide to Army Real Property Category Codes.* 11 Feb 00.
FM 7-0. *Training the Force.* 10 Oct 02.
FM 17-12-7. *Tank Gunnery Training Devices and Usage Strategies.* 1 May 00.
FM 17-12-8. *Light Cavalry Gunnery.* 16 Feb 99.
FM 100-14. *Risk Management,* Apr 98.
TC 25-1. *Training Land.* 30 Sep 91.
ARTEP 7-7J-MTP. *Mission Training Plan for the Mechanized Infantry Platoon and Squad,* 05 Oct 2001
ARTEP 7-8-DRILL. *Battle Drills for the Infantry Rifle Platoon and Squad.* 3 Mar 00.
ARTEP 7-8-MTP. *Mission Training Plan for the Infantry Rifle Platoon and Squad.* 05 Oct 01.
ARTEP 7-20-MTP. *Mission Training Plan for the Infantry Battalion.* 27 Nov 2001.
ARTEP 71-2-MTP. *Mission Training Plan for the Tank and Mechanized Infantry Battalion Task Force.* 27 Nov 2001.

**Section II**
**Required Documents**

These documents must be available to users of this publication.

AR 350-1. *Army Training and Education.* 9 Apr 03.
FM 7-1. *Battle Focused Training.* 6 Jun 03.
DA Form 2028. *Recommended Changes to Publications and Blank Forms.* Feb 74.
DA Form 2408-4. *Weapon Record Data.* Jan 79.

**Section III**
**Recommended Readings**

These publications contain relevant supplemental information.

AR 1-1. *Planning Programming and Budgeting and Execution System.* 13Aug 96.
AR 40-5. *Preventive Medicine.* 15 Oct 90.
AR 75-1. *Malfunctions Involving Ammunition and Explosives (RCS CSGLD-1961(MIN)),* 23 Apr, 01.
AR 95-2. *Air Traffic Control, Airspace, Airfields, Flight Activities and Navigation Aids.* 10 Aug 90.
AR 140-483. Army Reserve Land and Facilities Management. 30 Jul 94.
AR 200-1. *Environmental Protection and Enhancement.* 21 Feb 97.
AR 200-2. *Environmental Effects of Army Actions.* 23 Dec 88.
AR 200-3. *National Resources—Land, Forest, and Wildlife Management.* 28 Feb 95.
AR 200-4. *Cultural Resources Management.* 1 Oct 98

AR 350-2. *Opposing Force Program.* 15 Jun 83.
AR 385-10. *Army Safety Program.* 29 Feb 00.
AR 385-64. *Ammunition and Explosives Safety Standards.* 1 Mar 00.
AR 415-15. *Army Military Construction Program Development and Execution.* 9 Apr 98.
AR 415-35. *Minor Construction.* 15 Sep 83.
AR 420-74. *Natural Resources.* 10 Oct 97.
AR 700-127. *Integrated Logistical Support (ILS) Manager's Guide,* 11 Oct 99.
DA Pam 200-4. *Cultural Resources Management.* 1 Oct 98.
DA Pam 350-39. *Standards in Weapons Training (Special Operation Forces).* 3 Jul 97.
FM 3-04.140. *Helicopter Gunnery.* 14 Jul 03.
FM 3-20.12. *Tank Gunnery (Abrams),* 1 Oct 01.
FM 5-34. *Engineer Field Data.* 30 Aug 99.
FM 5-102. *Countermobility.* 14 Mar 85.
FM 5-250. *Explosives and Demolitions.* 10 Mar 86.
FM 6-30. *Tactics, Techniques, and Procedures for Observed Fire.* 16 Jul 91.
FM 6-71. *Tactics, Techniques and Procedures for Fire Support for the Combined Arms Commander.* 20 Sep 94.
FM 7-90. *Tactical Employment of Mortars.* 9 Oct 92.
FM 9-15. *Explosive Ordnance Disposal Service and Unit Operations.* 8 May 96.
FM 20-32. *Mine/Countermine Operations.* 29 May 98.
FM 23-1. *Bradley Gunnery.* 13 Mar 96.
FM 23-8. *M14 and M14A1 Rifles and Rifle Marksmanship.* 15 Apr 74.
FM 23-9. *M16A1 Rifle and M16A2 Rifle Marksmanship.* 3 Jul 89.
FM 23-10. *Sniper Training.* 17 Aug 94.
FM 23-11. *90MM Recoilless Rifle, M67.* 6 Jul 65
FM 23-14. *M249 Light Machine Gun in the Automatic Rifle Role.* 26 Jul 94.
FM 23-24 *DRAGON Medium Antitank/Assault Weapon System.* 3 Apr 90.
FM 23-25. *Light Antiarmor Weapons*. 17 Aug 94.
FM 23-27. *MK 19, 40mm Grenade Machine Gun, MOD 3.* 27 Dec 88.
FM 23-30. *Grenades and Pyrotechnic Signals.* 27 Dec 88.
FM 23-31. *40mm Grenade Launcher M203.* 20 Sep 94.
FM 23-34. *TOW Weapon Systems.* 17 Aug 94.
FM 23-35. *Combat Training with Pistols and Revolvers.* 3 Oct 88.
FM 23-41. *Submachineguns, Caliber .45, M3 and M3A1.* 28 Jun 74.
FM 23-65. *Browning Machinegun, Caliber .50 HB, M2.* 23 Dec 02.
FM 23-67. *Machinegun, 7.62-mm, M60.* 29 Feb 84.
FM 23-90. *Mortars*. 9 Dec 02.
FM 23-91. *Mortar Gunnery.* 1 Mar 00.
FM 44-1-2. *Air Defense Artillery Reference Handbook.* 15 Jun 84.
FM 44-62. *Air Defense Artillery Automatic Weapon Gunnery.* 6 Aug 69.
FM 44-102. *Procedures for Ballistic Aerial Target System.* 30 May 74.
FM 90-10. *Military Operations on Urbanized Terrain.* 15 Aug 79.
FM 90-10-1. *An Infantryman's Guide to Combat in Built-up Areas.* 12 May 93.
MIL-HDBK-828A. *Laser Safety on Ranges and in Other Outdoors Areas.* 15 Apr 96
NG Pam 25-1. *Training Site General Information Summary.* 28 Apr 95.
NG Pam 415-10. *Army National Guard Facilities Construction.* 15 Aug 94.
NGR 415-5. *Military Construction, Army National Guard Facilities Construction. 30 Nov 90*
NGR 415-10. *Army National Guard Facilities Construction.* 15 Aug 94.
NGR 420-10. *Real Property Operations, Maintenance, and OMARNG Minor Construction, Army National Guard.* 15 Feb 90.

STP 7-11BCHM14-SM-TG. *Soldier's Manual, MOS 11B, 11C, 11H, and 11M, Infantry Skill Levels 1/2/3/4.* 30 Sep 88.
STP 7-11BCHM24-SM-TG. *Soldier's Manual and Trainer's Guide, MOS 11B, 11C, 11H, and 11M, Infantry Skill Levels 1/2/3/4.* 19 Jan 00.
STP 7-11C14-SM-TG. *Soldier's Manual and Trainer's Guide MOS 11C Indirect Fire Infantryman, Skill Levels 1/2/3/4.* 16 Aug 94.
STP 21-1-SMCT. *Soldier's Manual of Common Tasks Skill Level 1.* Aug 03.
TC 6-40. *Field Artillery, Manual Cannon Gunnery.* 27 Dec 88.
TC 23-14. *Sniper Training and Employment.* 14 Jun 89.
CEHNC 1110-1-xxx series.

# Index

**S**



**T**



**U**



**V**

**W**



**X-Y-Z**

Made in the USA
Monee, IL
04 November 2026

67383095R00085